# CATASTROPHE:
# PAST, PRESENT, FUTURE . . .
# INEVITABLE?

- Is humankind headed for extinction—and if so, by what means?
- Has man developed—at no small cost—the ability to equal the doomsday potential of Nature?
- Why did Einstein say that "God plays dice with the cosmos"?

From subatomic particles to entire galaxies, from the genes within our cells to the masses beneath the Earth's surface, Fred Warshofsky analyzes actual and possible disasters in physics, geology, biology, and the natural world—and comes up with startling, controversial conclusions about our tomorrows.

FRED WARSHOFSKY

THE SCIENCE OF CATASTROPHE

PUBLISHED BY POCKET BOOKS NEW YORK

POCKET BOOKS, a Simon & Schuster division of
GULF & WESTERN CORPORATION
1230 Avenue of the Americas, New York, N.Y. 10020



Published by arrangement with Reader's Digest Press
Library of Congress Catalog Card Number: 77-7580



ISBN: 0-671-82208-X

First Pocket Books printing January, 1979

10 9 8 7 6 5 4 3 2 1



Printed in the U.S.A.

# CONTENTS

# PREFACE

"Science is the observation of phenomena and the communication of results to others," the Danish physicist Niels Bohr once remarked. From their observations scientists can create laws and are able to predict the manner in which observed phenomena will behave. Eventually the observations and laws tend to evolve into something larger, a paradigm, or model, into which all future observations and laws must fit.

Toward the close of the eighteenth century, after observing the structure of the earth for many years, scientists began to develop laws that explained the formation of the earth. Those laws fitted a model that called for the creation of the earth by catastrophic means—earthquakes, volcanic eruptions, tidal waves and other shattering phenomena. Geology—the science of the earth—was based on catastrophe. Catastrophe created mountains and oceans. Catastrophe was God's means of creation, and it fitted neatly with church dogma, satisfied the religious tendencies of the scientists who propounded it and helped keep the general population frightened and obedient.

Catastrophism, as the paradigm was known, suited the needs of the culture and society of the time, but it ignored, in many cases, the truths that lay evident in the stones of the earth. And so catastrophism clung to outmoded, easily disproved ideas that made it less a science than a theological pejorative.

Catastrophism sought to compress the geologic record into the literal time zones of biblical chronology. "To do this," says Harvard University Professor of Geology Stephen Jay Gould, "the catastrophist imagined a profound discordance between past and present modes of change. The present may run slowly and gradually as waves and rivers do their work; the events of the past were abrupt and cataclysmic—for how else could they fit into a few thousand years? Mountains were raised in a day and canyons opened at once. Thus, the Lord interposed His will to break the rule of natural law and placed the past outside the sphere of scientific explanation."

By counting the biblical begats, James Ussher, the Irish Archbishop of Armagh, in the year 1650 stated unequivocally that the earth had been created exactly 4004 years before the birth of Christ. Archbishop Ussher's pronouncement was added to the Gospel by inserting his data into the margins of the Authorized Version of the Bible in 1701. Soon the marginalia attained the weight of Gospel, and the time of the creation become even more precise as a result of the calculations made by Dr. John Lightfoot, vice-chancellor of the University of Cambridge. Dr. Lightfoot "proved" that "heaven and earth, center and circumference, were created together, in the same instant, and clouds full of water." Moreover, "this work took place and man was created by the Trinity on the twenty-third of October, 4004 B.C., at nine o'clock in the morning."

To reconcile Ussher's and Lightfoot's calculations with Isaac Newton's laws of motion, scientists propounded the theory of catastrophism. The catastrophists took the Bible literally and required catastrophic events to fit the time frame of the Ussher-Lightfoot calculations.

Newton, it should be noted, took the story of Genesis just as literally but needed an ordered set of rules to make God believable. "For it became who created them," he wrote of the celestial bodies, "to set them in order. And if he did so, it's unphilosophical to seek for any other origin of the world, or to pretend that it might arise out of a chaos by the mere laws of nature; though being once formed it may continue by those laws for many ages."

The advantage of catastrophism was that while hewing closely to the Bible, both celestial mechanics and the geology of the earth could be explained. But even when catastrophism was most widely accepted there appeared scientists who questioned the catastrophic explanation of the earth's history.

Charles Lyell and Charles Darwin were the major figures in the overthrow of catastrophism. Their theories would replace it with a new paradigm—that change and creation are slow processes that are still going on and that these processes are readily understandable. But the processes did not fit the Ussher-Lightfoot timetable of creation, espoused by organized religion. The entire concept of uniformitarianism, as the new scientific theory was known, was very useful in breaking the stranglehold religion had held for so many years on men's thought.

The result was to benefit all science and to spark even more aggressive inquiry into the processes of nature than had been possible in the past. Another result was to provide people with a certain smugness where nature was concerned, a feeling that everything was explainable, predictable and avoidable.

But of course, it is not. Physicists, after years of hurling fantastic energies at atomic targets in hope of understanding the complex forces of nature, have produced hundreds of new particles and antiparticles, have theorized such entities as quarks and such states as charms, have added greatly to our knowledge—and confess to being more confused than ever. At the 1974 meeting of the American Physical Society, its president, Dr. Wolfgang Panofsky, director of the Stanford University Linear Accelerator Center, told the assembled physicists that the new findings resulting from their experiments have led to a "state of maximum confusion" in the world of physics.

In astronomy the carefully drawn laws of Newton and Einstein, which helped explain so much of the mechanics of the universe, may not apply when black holes, the superdense remains of dead stars, are encountered.

On earth, volcanoes still explode to create new islands, continents collide and build mountain ranges, storms devastate vast areas, earthquakes reshape the land, and ice ages drastically alter the types and numbers of life. And on the cosmic scale, violent explosions of unimaginable fury bring death to some stars and birth to others. The entire rhythm of the universe may be the product of ancient catastrophe set in motion by a violence unimaginable to the mind of man.

Catastrophe is an essential force in nature, not aberrational, but inevitable. Catastrophe is a creative event that we must accept as such. There is no antiscientism about such a philosophical acceptance, nor should this book be considered an attack on science. Rather, it hopes to offer a slightly different perspective on our understanding of the universe. With that viewpoint we intend to point out once again what the British biologist J. B. S. Haldane had learned at the close of a long and distinguished career:

"The universe is not only queerer than we suppose, but queerer than we *can* suppose."

# THE SCIENCE OF CATASTROPHE

# 1

# THE GREAT DICE PLAYER IN THE SKY

"I cannot believe that God plays dice with the cosmos," Albert Einstein once said. It is perhaps the most often quoted of all the Einsteinian statements, for in it he encapsulated all of science's efforts since Aristotle to impose order on what at times seems to be a decidedly disorderly universe.

Newton discovered the laws of gravity, the force of attraction that objects have for each other. Every body in the universe, from the smallest particle to the largest galaxy, exerts that force. The strength of the mutual pull depends on the amount of matter, or mass, in the objects and the distances that separate them. Gravity, according to Newton, keeps the stars, planets and other celestial objects in their orbits.

To explain further the continuous, orderly processes of the universe, Newton then invented calculus. But not even for Newton would the stars behave precisely according to his laws, and so in his masterwork, *Principia Mathematica,* he called on God to intervene from time to time to reset the clockwork of the heavens to its original state.

The problem, for Newton and everyone else who attempts to understand it, is that nature suffers from discontinuities. Stars are created, burn steadily for billions of years, and then some of them suddenly blaze into supernovas and collapse upon themselves until they shrink into white dwarfs and eventually pull space in over themselves to become black holes. This is an example of a lengthy but nonetheless discontinuous process. And there existed, until recently, no satisfactory theoretical model capable of explaining discontinuous processes such as the destruction of a star, the boiling of a liquid, the breaking of a wave or the discrimination of different tissues in an embryo. But what Newton did for order, by inventing calculus, a French mathematician named René Thom has done for disorder by inventing a theory of catastrophe.

Thom built his theory on the results of 300 years of research, begun by Newton, into the manner in which continuously changing causes produce continuously changing results.

Using differential topology, a special kind of geometry concerned with the ways in which surfaces can be twisted, deformed, pulled, stretched or otherwise bent from one shape into another, Thom constructed a model of catastrophe based on folded sheets of paper. In this we are lucky, for topology is so far out at the boundaries of modern mathematics that it also deals with such seemingly impossible forms as a surface with only one side. But Thom places any particular continuous process on a smooth sheet of paper which has one fold, or pleat, in it.

As the process, such as a star burning its nuclear fuel to produce heat and light, climbs along the paper to the upper part of the fold, it will reach a point where it falls catastrophically to the lower part, like a boulder falling off the top of a cliff. Once it strikes bottom, it may begin its slow climb all over again.

Each catastrophe is the product of causative agents, and in using topological models, the fold becomes a theoretical cause of catastrophe. Using terribly involved mathematics, supplemented by computer graphics, Thom has explained his theory of inevitable catastrophe to the satisfaction of other mathematicians.

Disorder is also implicit in one of the classic laws of physics. The second law of thermodynamics says that the natural state of matter is chaos and that all things will eventually run down and become random and disorganized. In purely physical terms, the second law is a means whereby nature spreads out the available energy of the universe.

"It can be shown," says that irrepressible popularizer of science Dr. Isaac Asimov, "that any process which evens out energy concentration also increases disorder. Therefore, this tendency to increase disorder in the universe with the free random motions of the particles making it up, is but another aspect of the Second Law and entropy can be considered a measure of the disorder present throughout the entire universe."

Asimov demonstrates his point by arranging nine people in a square of three rows of three. The chance of all nine people together taking one step forward, and thus retaining the square, is 1 in 262,144. But in nature there are uncounted trillions of atoms free to move in very many different directions. "If by some chance there were some sort of order imposed on the arrangement of atoms to begin with," Dr. Asimov concludes, "then any free random motion, any spontaneous change, would be bound to decrease that order, or to put it another way, increase disorder."

In the same way Newton's desire to explain a static and unchanging universe extending on into infinity was doomed because of his own laws of gravity. Einstein pointed this out when he tried to explain the workings of the universe according to his general theory of relativity. But the objections he had found in Newton's theories applied to his own. ". . . the field equations of gravity which I have championed hitherto," he wrote, "still need a slight modification, so that on the basis of the general theory of relativity those fundamental difficulties may be avoided which have been set forth . . . as confronting the Newtonian theory."

Einstein knew what Newton could not: that another force besides gravity had to be at work in the universe. That force is known as electromagnetism, a range of radiant energies that travel in waves. Some waves are long, and some are short. The length of each wave determines the type of energy it carries. The longest waves of the electromagnetic spectrum are radio waves. At the other end of the spectrum are the shortest waves, known as gamma rays. Between these two extremes are the waves that make up X rays, ultraviolet waves, visible light and infrared waves.

Einstein's theory of relativity then showed how gravity acted upon electromagnetism. Space no longer stretched into infinity, islands of Milky Way-like bodies set in a flat sea of "empty" space. Relativity showed a new picture of the universe where space is curved, the path of light is bent, and time is slowed.

With his theories of relativity, Einstein was continuing Newton's efforts to make the universe behave predictably. Despite the evidence that proves the theories of relativity correct, Einstein failed ultimately in his efforts to make God's dice roll sevens at every pass. To do this, Einstein sought to construct what he called a unified field theory, a set of laws that would embrace both electromagnetic and gravitational fields as manifestations of the same phenomena. He spent the last thirty-five years of his life in this attempt and failed—frustrated by his inability to make his conceptual mathematics conform to the physical universe.

"The mathematical conclusiveness of the theory cannot be opposed," he wrote. "The question of its physical validity, however, is still completely undecided. The reason for this is that comparison of calculated solutions with experiment entails field equations that cannot be formulated."

Today Einstein might be even more frustrated by the avalanche of evidence that is pouring in from outer space and from the minds of other theorists. The discovery of new celestial objects, most of which seem to be the remains of cosmic catastrophes, continue to underscore the point that nature is not immune to sudden, explosive change.

Our view of those changes has been greatly improved by space age technology, which lifts our telescopes above the murky atmosphere of the earth that blocks and absorbs much of the energy radiated by the stars. By receiving and recording the X rays and other energies generated by the stars, new branches of astronomy, named for the radiations they study, have been created. X-ray astronomy is a major new field that gathers most of its information from telescopes aboard earth-orbiting satellites.

The first of these, SAS-A for Small Astronomy Satellite-A, was launched on December 12, 1970, from San Marco Island off the coast of the African state of Kenya. Since December 12 happens to be Kenyan Independence Day and NASA leases San Marco Island from Kenya, the naming of SAS-A as the satellite Uhuru was considered a happy choice by all concerned. As the first satellite devoted exclusively to X-ray astronomy Uhuru did not disappoint. Before launch the catalogue of known X-ray sources numbered 36; now it has increased to more than 200. Uhuru and subsequent companions have expanded astronomy and changed our picture of the galaxy.

Does that mean that Einstein, Newton and all those before, since and in between were wrong in their efforts to seek harmony in nature, to find and define the laws that govern the universe? Certainly not. Rather, what is proposed is that a philosophical adjustment be made, an acknowledgment of what man has learned by grievous experience through the millennia—that physical laws are true only to a degree; that the exception to the rule is inevitable and not exceptional; that many things in heaven and earth may be the product of mishap as much as of design. Now Newton's clockwork machinery lies shattered by relativity and quantum theory, which in turn are being buffeted by the astonishing welter of information, in the form of X rays, radio waves, gamma rays, cosmic rays and neutron fluxes that are pouring into our instruments from outer space. All these spectral notations are made from the detritus of cosmic destruction, of catastrophes of such enormity they are no more comprehensible than is the God we are still trying to make believable. For what we are finding in space is a cosmology built on catastrophe, events that must now be incorporated in any attempt to chart a uniform progression for the universe. Consider the evidence of the beginning of the universe in what is now popularly known as the big bang. Virtually all astronomers and everyone else concerned with cosmology, the study of the beginning, are now convinced that the universe, at least the one we inhabit, began as the product of an incredible explosion. Such an explosion was a radical change, an enormous discontinuity from whatever state had existed before and is, therefore, by definition a catastrophe, the most stupendous, mind-blowing catastrophe of all time. After all, it's how things began.

"In the beginning," according to Genesis, "God created heaven and earth." The Bible doesn't pay attention to anything before the beginning, in a sense very logically, and the creation myths of most other cultures also start from a void, a darkness or some sort of metaphysical square one before which there exists simply nothing. The last time anyone questioned the concept of God's square one was sometime around the fourth or fifth century when St. Augustine of Hippo was being heckled by a crowd of the less than faithful. "What," they wanted to know, "was God doing before He created the world?"

"Creating Hell for those who ask that question," the bishop snapped, and rarely has it been raised since.

Things nonetheless did begin, and the explanation or, at least, the first of those that utilize the big bang was also provided by a cleric—the Abbé Georges Lemaitre, a Belgian priest. His idea was that originally all the matter of the universe had been present in one huge cosmic egg, which became so hot it exploded, driving material out to form up into galaxies composed of stars, comets, planets and other assorted heavenly chunks of matter.

Physicist George Gamow first explained the physics that make such a process possible. He postulated a giant glob of very dense matter which he dubbed ylem—the name Aristotle had given to the stuff from which *he* believed all matter had been made. Gamow's ylem was composed of a primordial mix of protons and electrons, which was forced by the enormous pressure of gravity to unite into neutrons—elementary particles with no electrical charge. Then the cosmic egg of ylem exploded, and the neutrons were blasted out and began to fall apart into the protons and electrons of which they had been originally formed. Despite the size of the void, so vast was the collection of primordial particles they kept banging into one another until forces known as strong interactions acted like a nuclear glue to bind them together. A proton and an electron together would merge to form a hydrogen atom—the primal element. After a bit more time bouncing about it would collide with still other protons, neutrons and electrons to form still other atomic elements. The collisions kept up, according to Gamow, until hydrogen, helium, lithium and, in fact, all 106 plus elements we know today were formed. Gamow calculated that the entire chemical composition of the universe was completed in about thirty minutes after the ylem-fueled fireball had exploded.

It was an elegant, beautifully simple theory that explained everything. But it wasn't totally correct. In fact, Gamow was right for only about 15 minutes. The half-life of a free-flying neutron is only 12.8 minutes, and then it starts to decay back into protons and form up with electrons to make hydrogen. The rest of the atom construction took place not over the next 15 minutes, but much later, perhaps 5 billion years later. This was the flaw in Gamow's theory. If all the elements in the universe had been formed in that first half hour after the bang, the chemical composition of the universe would be the same throughout. In fact, it is not. Gamow soon realized this when everybody began saying it, and so he decided that hydrogen and possibly helium had formed right after the big bang and that the rest of the elements in the universe were created much later, the product of thermonuclear reactions inside the stars.

The Lemaitre-Gamow idea of a big bang to start things off is a good one, but it doesn't explain the catastrophic mechanism to everyone's satisfaction. What is the detonator of the cosmic bomb? Early catastrophism would allow God to violate the laws of physics with yet another piece of violence, turning the big bang into a God bomb. This would also fit biblical creation within the time frame of Gamow's half hour of chemistry, but we have already seen the flaw in that time scale. A detonator capable of setting off the big bang and fitting within the framework of the laws of physics is, however, possible.

The detonator is built from the particles that collected just before the big bang went bang. Gamow pictured this material as ylem. Another view of that primordial firecracker was made possible even earlier, when a British mathematician named Paul Dirac mated Einstein's special relativity to quantum mechanics and used the combination to describe the behavior of particles, the elementary components of the atom.

When man first peered into matter at the atomic level, he found in its minuscule makeup surprising parallels to the vast expanse of the universe. At the core of the atom was the nucleus, made of protons and neutrons. Around the nucleus were electrons whirling in what seemed to be orbital paths similar to those of the planets in the solar system. But there remained a number of mysteries that had to be solved. One involved the electron, for it did not follow a precise orbit and behaved at times in confusing ways.

In 1928 Dirac amplified the work of others to explain the motion of the electron. He showed that electrons were discrete bits of matter but that they behaved more like waves of energy. However, there was a disturbing corollary to his mathematics. The same equations led to the equally sound conclusion that an electron might possess not only a negative electrical charge (a perfectly natural state of events), but also a positive one (a decidedly unfamiliar affair). This would then mean that any nucleus surrounded by a positive electron would have to be negatively charged. In short, the combination of positive electron and negative nucleus would be the exact antithesis of atoms as we know them—an antiatom.

The idea was preposterous to most scientists, and even Dirac admitted his unhappiness with the results; but no one could challenge the mathematics. It was an excruciating dilemma, for the mathematics was truly beautiful—to a mathematician, that is—and one side of the equation, that of the negative or familiarly charged electron, could be precisely measured and confirmed. Only the flabbergasting positive electron could be neither observed nor measured; it merely existed as an affront to physical theory.

There matters stood for four years—an obvious impossibility made possible by the unshakable authority of mathematics. Then, in 1932, Dirac's equation was upheld by Dr. Carl Anderson, a physicist at the California Institute of Technology. Anderson was studying cosmic rays, charged particles that bombard the earth from outer space. Cosmic rays enter our atmosphere at fantastic velocities, collide with its atoms and shatter them into the detritus we know as subatomic particles. To record the nature of these particles, Anderson used a cloud chamber—a container filled with an alcohol-saturated gas. A particle entering the chamber leaves a visible trail of droplets in the gas. A magnetic field set up in the chamber will push the particle along a curved path. From this curving necklace of droplets, the speed, mass and charge of the particle can be measured.

In the course of photographing the track of the cosmic rays one day, Anderson made a startling discovery: A particle with the same characteristics of an electron had left its highly revealing footprints in the cloud chamber. But this was like no electron track ever before seen, for it curved not toward the right, as any other electron would have done, but rather to the left. The meaning was unquestionable: This was photographic evidence of an electron with a positive charge. Anderson dubbed it the positron, for it was the same oddly behaved positive electron that Dirac had predicted.

Later the development of atom-smashing cyclotrons and

accelerators allowed scientists to bring their search for antimatter particles into the laboratory. There, in 1954, they found the antiproton, and then, in 1965, the first complete nucleus of antimatter was found by Columbia University physicist Dr. Leon Lederman. This meant that the building block of matter, the atom, could also exist as an antiatom.

"A literal antiworld populated by stars and planets and made up of atoms of antimatter may well exist physically in addition to the known material universe," marveled Lederman after he had created the antinucleus.

The entire concept of antimatter so intrigued a pair of Swedish cosmologists—Hannes Alfvén and Otto Klein—that they proposed a detonator for the big bang, composed of both matter and antimatter. For when the two combine, there is a catastrophic orgy of mutual annihilation—as matter and antimatter together are completely transformed into energy. The phenomenon had already been predicted by Einstein in the theory of relativity, and the exact amount of energy to be released was contained in his famed equation $E = mc^2$. The equation states that matter is simply congealed energy and can be transformed back into energy with no loss of efficiency. In the positron, Einstein had demonstrable proof of his formula, for the positron was doomed to extinction the instant it met its opposite number, the electron. The two antithetical particles indulged in an orgy of mutual annihilation and released fantastic energies in the process.

And so the Swedish version of the big bang starts with a mixture of matter and antimatter Alfvén and Klein call ambiplasma, a term derived from the Latin *ambo*, meaning "both."

The two Swedes think big, visualizing at first a spherical cloud of ambiplasma with a radius of perhaps a trillion light-years. In it nothing would happen because particles and antiparticles would rarely meet and annihilate each other, for each would have about a million cubic meters of empty space to rattle around in. But if gravity began to pull the cloud into a diameter of only about a billion light-years, the number of particle-antiparticle collisions would increase, increasing the amount of radiation in the system.

Then the two forces, radiation and gravity, would combine both to pull the cloud still more tightly together and to agitate the particles even more. Finally, the contraction would reach a point where gravitational forces would be overwhelmed and the mass would explode, detonating the big bang.

Matter-antimatter annihilation is a neat explanation for a big bang, but it doesn't explain why the annihilations did not continue until nothing was left but raw energy. The reason, says Professor Alfvén, can be seen on a hot stove. Here we can find a demonstration of the Leydenfrost phenomenon. If water is dropped on a very hot, flat plate, it boils and evaporates very quickly. But if the plate is concave and can keep the water from rolling off, a teaspoonful can remain without evaporating for as long as ten minutes. The reason is that a film of water vapor forms between the plate and the water, thereby insulating the water from the heat of the plate.

According to Alfvén's mathematics, a similar Leydenfrost insulation is created by a thin, very hot layer of ambiplasma that prevents stars and even galaxies from annihilating each other.

Antimatter is a delicious idea; it obeys the laws of symmetry that physicists demand. It offers God a means of detonating His bomb and can even, by some fanatical stretch of the theological imagination, be seen as the byproduct of the war between good and evil, between God and Satan. Unfortunately antimatter cannot be seen in space since stars look very much the same and do not reveal their matter or antimatter pedigrees in their spectral signatures. "It cannot be said for sure that the Andromeda galaxy or even our own galaxy does not contain antimatter," Dr. Klein notes sadly.

As with all the other cosmologies, matter-antimatter remains theoretical, a victim of observational failure. Ultimately, astronomers and physicists insist on observational evidence to support theories. Where cosmology is concerned, one of the few seemingly meaningful observations that have ever been made occurs in what astronomers call the red shift.

Since 1888 astronomers have known that as stars moved away from the earth, their light would be shifted toward the red end of the electromagnetic spectrum. When receding, the visible light spectrum is bounded on one end by a violet band and on the other by a red one. The faster the flight of an emitting object, the more its light is stretched out toward the red end of the spectrum, where the waves are longer. The faster it moves away, the bigger is this so-called red shift.

In 1929 Edwin Hubble, an astronomer at the Mount Wilson Observatory, measured the red shifts of several galaxies and found they were extremely pronounced and were therefore moving away from the earth at an extremely rapid rate. He then put forward the idea that this speed was also a measure of distance and that all the observable galaxies were moving away from each other.

Two conclusions were apparent: One, the universe was expanding and two, the cause of that expansion was very probably a great explosion—a big bang.

The reasoning behind the conclusions was simple. In effect, Hubble viewed the movement of the galaxies away from one another as recorded on motion-picture film. But he came in somewhat late in the movie, after it had begun. To see it from the beginning, he simply ran the film backward. This reversed the outward rush of the galaxies and sent them hurtling in toward some common starting point. As they drew closer together, the density of the material increased until all the matter and energy of the universe were packed into an incredibly dense glob. That moment was time zero, when Gamow's ylem reached a critical point and went boom.

According to Hubble, the time at which the big bang occurred could be determined by measuring the speeds at which the galaxies are receding from one another today. Hubble then devised a mathematical constant that made the speed of the recession proportional to the distance of the galaxies. Using this constant, Hubble determined the age of the universe at 1.8 billion years.

The Hubble constant made the universe considerably older than the age given by Archbishop Ussher, although less precise than the starting time calculated by Dr. Lightfoot. Then the Hubble figure was found to be equally incorrect. The ages of the rocks in the earth's crust were known to be older than the age Hubble had ascribed to the universe; this was, of course, impossible.

The apparent youth of the universe led a trio of Englishmen, Herman Bondi, Thomas Gold and Fred Hoyle, to develop an alternate theory that would account for the headlong rush of the galaxies away from each other by a less catastrophic mechanism than the big bang. Bondi,

Gold and Hoyle postulated that fresh new hydrogen, the basic stuff of the universe, was being continuously created out of nothing. Moreover, they said this hydrogen production goes on throughout the universe at a rate that exactly fills in the gaps left by the receding galaxies. The new hydrogen also provides the building materials from which new galaxies are formed, thereby maintaining the average distance between them.

The theory was called the steady state universe. If correct, the steady state theory meant that the universe was unchanging and its age infinite. The theory was a mathematically possible alternative to the big bang. As long as observational evidence was lacking, either the big bang or the steady state could serve as the proper model of the universe. Both theories accounted, in radically different ways, for the only observed phenomenon that had cosmological significance—the receding galaxies.

Then a big banger, a Princeton University physicist named Robert Dicke, decided, as had Gamow, that there should be some cosmic residue, a sort of fireball radiation left over from the big bang. The big-bang afterglow is known technically as black body radiation, and the appropriate frequency ranges in which it might be found in space were then computed by Dicke and his colleagues.

No sooner had the calculations been published than a pair of radio astronomers at Bell Laboratories consulted Dicke and realized that the strange radio signals they had been receiving from outer space were, in fact, the dying embers of the big bang.

The discovery of black body radiation flooding in from all points makes the universe a sort of gigantic microwave oven from which no radiation can escape. Moreover, the temperature inside that oven is much colder than it was in the past. "This," according to NASA astrophysicist Floyd Stecker, "indicates that at some time in the past the universe must have been at a state dense enough to reach thermal equilibrium [when all parts were the same temperature], as indicated by the black body frequency distribution of the radiation. For thermal equilibrium to be attained, the universe would have once had to have been at least a billion times more dense than it is at present and hot enough so that all of the matter in the universe was in the form of an ionized gas, or plasma. This fact is what has convinced almost all astronomers to accept big bang cosmology as opposed to steady-state cosmology, which says that the universe was always at the same temperature and density as it is now."

So it's back to square one, but there remains that annoying problem of *when* the big bang occurred, for Hubble was obviously wrong about the moment of creation. What was at fault, however, was Hubble's constant, not his formula, which was considered a far more reliable method of computation than counting the begats of the Bible. So over the years Hubble's constant has been constantly revised. The latest revision places the farthest galaxies even farther away than Hubble ever dreamed, and the adjustments to his constant make the universe older still. The man chiefly responsible for the continuing revisions is Dr. Allan Sandage, of the Hale Observatories. His latest revision of Hubble's constant, made in 1972, now gives an age of 17.7 billion years to the universe. The Sandage figures also indicate that the rate of expansion is slowing, but not enough to reverse the process. This news came as something of a blow to the philosophers and theologians who see in the big bang the hand of the creator.

Any supernatural being must by definition be not only omnipotent, but infinite in terms of his/her/its existence. In the concept of a closed or oscillating universe that expands and contracts, forever producing big bangs ad infinitum, the immortality of the deity is maintained.

The alternative, with all its nihilistic implications, is that the universe is open, that it will continue to expand until all the stars die out and all life everywhere ceases to be in a cold, black, empty void.

So more than a century after the demise of catastrophism we find the theologians, who originally needed catastrophe to explain the creation as God's work, united with the cosmologists, scientists who cling to a catastrophic big bang as the one event capable of providing the mechanics of creation that follows the physical laws they have drawn.

The amount of matter left over in space after the creation of the universe is the key to whether the universe is closed or open. The known mass of the universe in the form of stars, planets and other large bodies found in the clumps we call galaxies does not add up to enough matter to act as the gravitational brake needed to slow expansion to the point where it will begin to contract. The hope has often been expressed that the required material would be found in intergalactic space, in the form of huge clouds of material too dim to be seen but heavy enough in the aggregate to provide the needed mass to keep the universe from forever flying apart. Dr. Jeremiah Ostriker of Princeton theorized that the galaxies are surrounded by huge spherical halos of stars and other materials. These invisible halos, he proposed, are ten to twenty times as large as the galaxies themselves and increase their mass tenfold.

Another astrophysicist, Dr. James Gunn of Cal Tech, agrees with the Ostriker idea and says that "possibly most of the mass of the universe resides in this silent majority of small stars."

Theoretically, there should be about thirty times more mass in the universe than has been found by present measurements. That missing matter would provide the gravitational brake needed to slow expansion and ultimately to contract the universe back to a point where the big bang can again take place. But if the silent majority of stars increases the amount of matter by tenfold, we are still twenty times short of the amount needed to brake expansion. And so the search began for the "missing matter of the universe."

On August 21, 1972, a 4900-pound $83-million satellite named Copernicus was launched into space. Among its tools NASA pointed with pride to its 32-inch ultraviolet telescope capable of sighting a basketball on the earth's surface from its orbit 425 miles in space. The telescope, of course, was not looking for basketballs, nor was it even pointed at the earth. It was aimed out toward deep space, where, unimpeded by the atmosphere, it was looking for the missing matter in the universe. It didn't find it. Instead, like a Nielsen rating, it counted a random sample of atoms and found a mere $1.5 \times 10^{-31}$ grams of matter per centimeter cubed. That figures out to about one minute particle in every 60 feet of extragalactic space. This was a disheartening thirty times less than the minimum of $3 \times 10^{-30}$ grams per centimeter cubed needed to stop expansion and send the universe crashing back in upon itself so that another big bang could take place.

Although virtually every cosmologist concerned cautions that this is not the final answer and the figures are not absolutely accurate, the evidence indicates that the final catastrophe will be a demise much like that prophesied by T. S. Eliot—"not with a bang, but a whimper," a whimper that will probably be immediately swallowed by the most colossal finding in space to date, the black hole.

The black hole is an astronomical phenomenon of such enormity and such incredibly bizarre characteristics that it is now being touted by some cosmologists as a creative alternate to the big bang. But what exactly is a black hole? Simply stated, it is a dying star which has collapsed upon itself. Neither light nor matter can escape because of its intense gravitational field. As a theoretical concept the black hole is probably no more difficult to come to grips with than the big bang, but the observational possibilities are decidedly more limited. For by definition the black hole emits neither light nor matter, hence should be undetectable, while the dying embers of the big bang have been detected, and the rushing surge of galaxies away from one another, another evidentiary point, is also observable. But the black hole remains mute, impenetrable. Yet black holes, according to theory, start out as perfectly ordinary stars, much like our sun. The sun formed some 5 billion years ago when huge masses of hydrogen were pulled by gravity into a sphere. The pressure and temperature at the center of the sphere built up to the point where the star ignited its thermonuclear furnace. Protons smashed against protons, forming helium and then still heavier elements. In the process great amounts of energy were liberated and in our case eventually bathed the earth with enough heat and light to make it a fit place to live.

The process, however, is a delicate balancing act whereby gravity pulls the star inward—just enough to keep the nuclear furnace going. Energy from the nuclear reaction pushes outward—just enough to keep the star's gravitational forces from forcing the star to collapse upon itself.

At some stage in the sun's lifetime, however, the nuclear furnace will exhaust its store of hydrogen fuel and begin to swell, blowing off gases from its outer shell. Soon the sun will expand to a red giant, growing to some 250 times its present diameter of 850,000 miles and in the process devouring Mercury, Venus and the earth.

Soon, however, as the remnants of the nuclear fuel are used up, the bloated sun will begin to contract down through its present diameter to a mere one-hundredth of its present size until it is just about the size of the earth—but 210,000 times as dense. A Ping-Pong ball of its material would weigh as much as a herd of elephants. At this stage the star is called a white dwarf, and its contraction stops. The electrons of its atoms are so tightly packed together that they create a pressure effective enough to counteract the gravitational contraction that wants to pull it even more tightly together. And here the history of our sun will end, perhaps in another 5 billion years; but the mass of our sun is less than that of most stars, and the lives of these larger ones can end far differently.

If a star contains only twice as much matter as our sun, it will not settle down into the dense, dead cinder that is a white dwarf. Rather, the additional matter will build such extreme pressures and densities that the packed electrons will be unable to generate a sufficient counterpressure. The star will rush past the white dwarf stage to explode catastrophically. That catastrophic explosion is called a supernova, possibly the most spectacular release of energy the universe can manage.

Supernovas have been seen exploding in other galaxies and even in our own. The Chinese recorded one in July 1054, thought it a newly discovered star and named it the Guest Star. Within a week their new star was so bright it could be seen in broad daylight. Then, just as quickly, the Guest Star faded from view, and by the end of the year it was gone forever. What remained, however, was a glowing fingerprint in the sky, a huge blur of gas we now call the Crab nebula. The star had actually exploded 4100 years before the Chinese saw it, for that is how long it took the light from the explosion to reach us from the star that lay 246 quadrillion miles away from the earth. The Crab nebula is still expanding, at a rate of 800 miles a second, and it now stretches across some 35 trillion miles of space. It is a colossal neon headstone marking the death of a star, but it is not the final end.

Possibly as much as 90 percent of the Guest Star's mass has been blown away by the colossal supernova explosion, leaving only the collapsed core of the star at the center of the still rapidly expanding Crab nebula. The core is much smaller than a white dwarf and must find some form of equilibrium so that it can once again be stabilized. It does so by packing its protons and electrons so tightly together they are squeezed into neutrons. Then the neutrons are squeezed still further. The result is an incredibly dense neutron star which could contain all the mass of the sun in a sphere only 10 miles across. On the surface of the neutron star gravity would be 210 billion times that of the earth. One cubic inch of the neutron star would weigh a billion tons.

Moreover, neutron stars spin at phenomenally rapid rates, spewing forth radio signals with better than metronomic precision. As a result of these radio pulses, the neutron stars are better known as pulsars. "The bursts of radio waves from these sources," Jeremiah Ostriker points out, "were so accurately timed that any of the four originally discovered pulsars could have been used as a clock accurate to one part per 100 million."

The "ticking" of the pulsars was so rapid and constant, ranging from a quarter of a second to 1.3 seconds, they gave rise to stories about little green men attempting to contact the earth. "Our first thought," said Sir Martin Ryle, a British astronomer, "was that this was another intelligence trying to contact us."

The discovery of many other pulsars and the realization that the pulses were not communications from little green men but were proof of the neutron star concept, which had first been proposed by Robert Oppenheimer in 1939 and greeted with laughter by the astronomical community. With the discovery of pulsars, now the ideas of other theoretical astrophysicists took on a new urgency. They had postulated that under certain conditions even the seemingly absolute resistance of the neutron structure can be overcome by gravity. Then there remains nothing to resist the ultimate collapse. The star can shrivel to zero volume, and the surface gravity can zoom toward the infinite.

At this point Newtonian laws of gravity no longer apply, and physicists turn for assistance to Einstein's general theory of relativity. What happens then, in the curved space-time of Einsteinian mathematics, is so bizarre, so extreme, that everything else in nature must forever after appear commonplace. "Assuming the exact spherical symmetry is maintained right down to the center," explains Roger Penrose of London University, "the answer provided by the general theory of relativity is a dramatic one. Not only is the substance of the original body squeezed to infinite density at the center of the black hole—that is, effectively crushed out of existence—but also the vacuum of space-time outside the body becomes infinitely curved. The effect of this infinite curvature on a hapless observer, were he foolhardy enough to follow the body inward, would be catastrophic. He would feel tidal forces across his body that would mount rapidly and would reach infinity within a finite period of his experienced time."

And no one would see what was happening, for not even light can escape the incredible tidal forces of gravity that pull it inward. It's as if a trapped object had fallen into an infinitely deep hole and never stopped falling.

If the universe is riddled with those black holes and neither light nor sound nor any other form of energy can escape from them, they should be impossible to detect. Or are they? Some years ago a pair of Russian theorists put forward the idea that if a black hole were orbiting a larger, visible star, it would draw gases from the star. As those gases spiraled into the black hole, they would collide, compress and heat up to a temperature that could reach 100 million degrees—hot enough to generate an intense stream of X rays.

Among the most prodigious X-ray sources in the universe is an invisible member of the constellation Cygnus, the Swan, about 6000 light-years away. The X rays, representing a million times more energy than the total output of our sun, were first picked up by the Uhuru satellite and dubbed Cygnus X-1. At first Cygnus X-1 was thought to be a pulsar, but its wildly fluctuating streams of X rays were very different from the steady bursts of radiation scientists have come to expect from pulsars.

The X-ray flood was so powerful and so dissimilar that radio astronomers all over the world began to focus their antennas on the radio signals coming from the area. The signals provided a much more precise fix than Uhuru had given, and at last the astronomers were able to train their big optical telescopes on Cygnus X-1. There they uncovered a massive star, a so-called Class-B supergiant, at least twenty times larger than our sun. It was weaving drunkenly through space, following an erratic orbit that was obviously unnatural. Something, a smaller companion star, was tugging it off its more normal orbit.

But having pinpointed it in space, a team of British scientists were able to orient NASA's Copernicus satellite on an invisible star—the cause of the supergiant's erratic orbit. By observing the way X rays are absorbed as they pass through the visible star's atmosphere, the astronomers have concluded that the X-ray object is very small. The mass of the invisible star is more than three times that of our sun, but less than one-fiftieth of its size. Still, its gravitational pull is so great that it affects the orbit of the supergiant it circles. Such an invisible star can be only one thing, according to Dr. Peter Sanford, who did the direct observations and reductions of the Copernicus data: "It's a black hole."

Now that we have actually found a black hole the bizarre nature of its characteristics becomes even more frightening. Within the black hole there may lie the final cosmic obscenity, "the place," as one science writer put it, "where 300 years of physics gets the shaft."

The problem is that at the heart of a black hole, where it begins to approach the ultimate, there lies a "singularity." The term is actually a euphemism for bizarre events in mathematics and physics that seemingly follow no laws at all. To a mathematician a singularity is the point where infinities and discontinuities suddenly show up in otherwise placid equations. To a physicist it is the particle jungle where the laws of physics no longer apply. To a cosmologist who uses general relativity it is the place where space-time becomes twisted completely out of shape.

The deeper you go inside a black hole, the sharper the curve of space-time is twisted until, at the center, the twist is so sharp that space-time disappears. You have reached a singularity. Thus, just as Newton's laws collapse under the extreme conditions where relativity becomes dominant, so relativity crumbles as it falls deeper into the black hole, where conditions are so extreme that the laws that govern them are totally unknown.

Cosmologists have seized on the black hole as a means of explaining the unexplainable. The case of the missing matter of the universe, for example, is solved, according to some scientists, by having enough black holes to provide the needed material.

Thus, from any black holes left over from the time of the big bang there will be enough mass to brake expansion, and the universe will once again collapse upon itself for yet another big bang. Kip Thorne, a theorist at the California Institute of Technology, explains that we are within a universe composed of space and time created by the catastrophic event we call the big bang, "and we are trapped inside its gravitational radius. No light can escape from the universe."

This may be the ultimate catastrophe to be suffered by

man's ego. For if no light can escape from our universe, how else can anyone in any other universe know we are here?

Before we despair completely, we may be comforted by the thought that our black hole will ultimately poke through into another universe, there to appear as a blazing "white hole." At least this is a thought of Yuval Ne'eman of Israel. He theorizes that quasars, those incredibly distant but extremely powerful stellar objects found at the borders of our visible optical telescopic vision, too far to be ordinary stars and still be seen on earth, are actually the remnants of material swallowed by a black hole which is being dumped into our universe from somewhere else.

While using the strange events within the black hole to power such speculations, theoretical physicists have managed to clamp a lid on the heart of the black hole, the singularity. It's almost a Catch-22, for every singularity has a black hole and every black hole a singularity. But they both are cut off from the universe by something called the event horizon. The event horizon is the surface of the black hole where gravity is so strong that matter cannot escape and light is totally extinguished. Communication across the event horizon is impossible. The physicist outside the black hole cannot get any information from inside it and has no way to understand the laws which govern it. Without that understanding he need not seek the laws since they are impossible to understand. Catch-22 again.

But a British physicist named Stephen Hawking threatens to upset this delicately balanced structure by saying that some types of communication are possible across the event horizon. A transfer of energy or matter could lead to the explosion or evaporation of the black hole. But according to general relativity, black holes should last forever, hiding their workings in the cloak of singularity. And without singularity, quantum mechanics and relativity become less than accurate predictors of physical events.

Thus, nature, on both the cosmic and subatomic level, reaches a point where it no longer follows the rules. Physical events that cannot be predicted take place, and catastrophe cannot be overlooked. All the efforts of Newton, Einstein and the others to keep the dice out of God's hands are undone. Not only does God play dice with the cosmos, but, adds Stephen Hawking, "He sometimes throws the dice where they cannot be seen."

# 2

# FAILED STARS AND EXPLODING PLANETS

First the good news. The death of a star is inevitably accompanied by the birth of another. Within the universe, birth and death are almost simultaneous events. Stars explode with cataclysmic fury, blowing off clouds of gases and molecules of matter across great regions of a galaxy. After a time gravity begins to pull the cloud back together. Eventually the mass becomes dense enough to ignite the nuclear furnace, fusion takes place, and a new star is blazing its light through the heavens.

Now the bad news. Our sun is dying. Most solar physicists had assumed our sun to be about 5 billion years old and gave it another 5 billion years of life on the main sequence—that is, as an effective thermonuclear furnace. But a recently concluded experiment indicates that the sun may already have gone off the main sequence. That hesitant conclusion is the result of work done 4850 feet beneath the surface of the earth in a rock cavern in South Dakota. Dr. Raymond Davis of the Brookhaven National Laboratories on Long Island for the better part of five years has been underground, seeking an incredibly elusive particle called a neutrino. Neutrinos are the product of beta decay, the process that breaks down neutrons into protons, electrons and neutrinos during a certain type of nuclear fusion. The neutrino is remarkably elusive; possessing no mass and no charge, it travels at the speed of light and passes through virtually any solid matter in its path. It should, in the normal course of cosmic events, reach the earth and pass right on through it about eight minutes after being produced in the sun's nuclear stewpot.

Other particles zip through the earth's atmosphere with almost as much abandon as the neutrinos, and it was to shield them from cosmic rays that Davis took his experimental equipment deep underground in a mine near Lead, South Dakota. For the neutrino was not completely undetectable. Earlier experiments had shown that on rare occasions, say one of every 1,000,000,000,000,000,000,000 neutrinos streaming through a very large tank of water will crash into a proton and interact with it. That interaction, of course, can be detected, and an indication of the actual number of neutrinos streaming out of a nuclear source can be gained. The initial water tank experiments were done with neutrinos cast off by a nuclear power plant. To catch a neutrino from the sun, Davis used the earth itself, or at least that 4850-foot-thick portion of it above his head in South Dakota.

The five years of neutrino catching yielded no more of the little neutral ones (the name Enrico Fermi gave them) than could be explained very easily as coming from sources other than the sun. The conclusion: The sun is not emitting neutrinos. The implication: The sun is dying, for without neutrino emission there is no evidence of thermonuclear reactions taking place.

Does that mean the white-sheeted sign carriers are right and the world will indeed end in a week and a half? By some theoretical reckoning, the sun will collapse just fourteen minutes after the nuclear reactions cease within its core. Another construct says that the energy the sun produces at its core takes from 10 to 50 million years to percolate out to its surface before it blows out over the rest of the solar system, thus removing the threat of momentary extinction.

Most solar physicists, however, consider reports of the sun's death, as did Mark Twain of his own, "greatly exaggerated." Not that they discount the Davis experiment. Rather, some scientists now believe the nuclear furnace of the sun may not burn constantly, but cycle, much like an air conditioner, in response to conditions in the surrounding environment.

Others see in the Davis experiment a crisis in solar physics. Conceivably it could lead to a totally new understanding of the way in which the sun produces its prodigious energies. In the past, experiments that delivered totally unexpected results have either been ignored or have led to profound changes in fundamental ideas. For the moment, the sun and the rest of the solar system that depends on it are safe, but they are not, as most people throughout our history have thought, unchanging and immutable.

The Greeks saw the earth as the center of the universe, circled by seven planets—the six they could see, plus the sun, which they considered merely another giant in the train that circled the earth. Wheels within wheels, all neatly arranged to fit an emotional conception of the heavens. And even after Copernicus had demonstrated that the sun, not the earth, was the center of the solar system and after Kepler had mathematically proved that the ellipse, rather than the perfect circle of the Greeks, was the path followed by the planets about the sun, it had little meaning to a church-oriented world that insisted that the view of Aristotle and Ptolemy was the only acceptable view of the heavens.

But the assaults on order and neatness continued when Galileo Galilei put his eye to the telescope and said in wonderment, "I have seen stars in myriads which have never been seen before, and which surpass the old, previously known number more than ten times. But that which will excite the greatest astonishment by far, and which indeed especially moved me to call the attention of all astronomers and philosophers, is this, namely, that I have discovered four planets, neither known nor observed by any one of the astronomers before my time."

The report was indeed astonishing, but not acceptable, for it constituted a heresy against the established order of things. The heavens were not mutable, doctrine and dogma were exact and unchanging, and so there could not be four more planets. Indeed, Galileo's four new planets were in reality the moons of Jupiter. One Florentine astronomer swiftly did away with Galileo's four new planets by arguing not that they were Jupiter's moons, but that as there were only seven openings in the head—two eyes, two ears, two nostrils and one mouth—and only seven metals and seven days in the week, so there could be only seven planets.

In looking backward at that time, the Florentine's reasoning for a solar system of seven planets is every bit as good as that for nine. In fact, it was not until 1930 that the ninth planet, Pluto, was discovered, and even now there are serious doubts whether it is a true planet or an escaped moon of Neptune's. There is no special reason for nine, or eight or two for that matter, for the question depends on the way in which the planets were created and the amount of material available with which to create them. There are several theories that explain planetary formation, one of which is heavily dependent on a rather shocking catastrophe. It says that our sun was sideswiped by a passing star. Huge gouts of gas were torn from both stars to flame across a vast region of space. Then, as the invading star retreated into the void, the gravitational attraction of the sun pulled the huge flaming streamers toward it; like taffy stretching, the streamers broke and then coalesced into balls of flaming gas held in thrall by the sun's gravity. Slowly the gases cooled to form a molten mass, were crusted over and hardened over millions of years to form the planets.

There are a number of holes in the star collision idea, not the least of which is the fact that the chances of a collision between the sun and another star are, well, astronomic. The sun's nearest neighbor is 26 trillion miles away, and the possibility of collision is, as we have said, remote. Moreover, if the collision theory were correct—and from our studies of space we see no evidence of collisions between stars—it would mean that very few planetary systems other than our own existed, and we know this is not the case by a factor of several hundred million.

The more likely process seems explained by the physics of gases. Observations have shown that gases in space are possessed of complex, internal, swirling movements. Thus, any portion of gas that condenses into a star must start spinning. The rate of spin is controlled by the density of the material. Just as an ice skater will spin faster as she draws her arms into her body and her skates closer together, so a condensing cloud of gas will spin more rapidly as the cloud grows more compact. The physical law that describes this process is known as the conservation of angular momentum. It also says that our sun should be spinning at a much faster rate than the once-every-twenty-seven days rotation it actually possesses. Something is obviously acting as a brake or drogue chute on the sun to slow it down. Moreover, that gravitational brake must have been in operation early in the sun's history, while it was still forming, or else it would, by our own rules of physics, be spinning faster than it actually does.

The brake appears to have been the planetary material shed by the condensing sun while it was still spinning rapidly. "This led to an outer ring of planetary material that surrounded the sun," explains British astronomer Fred Hoyle. "Then, an action-reaction was set up between the planetary material and the sun, slowing the sun's spin and, at the same time, forcing the planetary material further and further outward—again because of the principle of Conservation of Angular Momentum. The planetary material cooled as it went, so that various substances began to condense within it, initially as smoke and as droplets. Those substances that resist evaporation—rock and such common metals as iron—were the first to condense, and out of these materials were formed the innermost planets, including the earth. As the gases continued their outward journey, the temperature continued to lower, and less refractory materials, such as water, began to condense. These formed the outer planets."

Hoyle's picture of a solar system filled with planets that condensed out of the sun's leftover gases has recently received some support when Pioneer 10 flew past Jupiter, the largest planet in the system. Based on Pioneer's reporting, Jupiter appears to be a remnant of the earliest chemistry of the solar system. With a mass that is almost two and a half times greater than that of the other planets put together, Jupiter is composed primarily of the two lightest elements, hydrogen and helium. Its great size and composition of primal elements led some astronomers to suggest that Jupiter was a stillborn star, a mighty cloud of condensing gases that failed to achieve the size and temperature required to ignite a nuclear furnace. In many other respects it can be viewed as a solar system in miniature, with its four "Galilean" moons revolving about Jupiter and displaying the same heavier composition as the four inner planets of the sun. The other four moons of Jupiter contain the lighter elements, and it is thought that just as the sun's heat initially drove off the lighter elements from the inner planets, so too did Jupiter, which radiates two and a half times more heat than it receives from the sun, despite its failure to become a star. The heat was enough to drive off the volatile lighter elements from Jupiter's inner moons and allow only the four outer satellites to be enriched with the lighter elements. The parallel is, of course, identical in the four gas giants that orbit the sun beyond Mars, the last of the inner or heavier planets.

The actual planet-making process may have taken but an eye blink of cosmic time, perhaps as little as a few thousand years, according to some planetologists. It was, however, a time of intense activity.

"During the detailed formation of the big planets," explains Professor Hoyle, "there must have been a great deal of dynamic activity. Vast swarms of smaller bodies must have been formed first. These bodies must have collided frequently and, as a result, an enormous spray must have developed everywhere throughout our solar system. Probably the comets are surviving pieces from that spray. The pieces consisted of icy bodies—some of ordinary ice, some of dry ice, others of more esoteric substances, such as hydrocyanic acid (which is thought by many chemists to have played a key role in the origin of life). They flew like great snowballs throughout the solar system, and many of them must have hit the newly formed earth. Besides scarring the surface, as the moon has been scarred by such impacts, these collisions added water and other life-forming substances to the earth. It is possible, perhaps likely, that we humans owe our existence to that early rain of vast snowballs."

Those cometary snowballs still fly with frightening abandon through our solar system, occasionally smashing into the planets, but now more likely shedding bits and pieces to fall through the earth's and other planetary atmospheres, usually burning up, occasionally impacting with incredible force. No one is certain how comets originate, and a lively debate can usually be started among astronomers who hew to one or another theory.

Comets, say the traditionalists, form in much the same fashion as stars. A cloud of interstellar gas and dust collapses and cools to the point where the hydrogen, nitrogen, carbon and oxygen in its head freeze. Their birthplace is, according to the Dutch astronomer Jan Oort, a vast cloud of debris that stretches from beyond Pluto halfway to the nearest star, Proxima Centauri. Within this vast swarm of debris, known as Oort's Cloud, there may be as many as

100 billion comets, and all are gravitationally part of the solar system. According to Oort, a star passing near the outer reaches of this cometary cloud will disturb the orbits of some comets just enough for them to enter the inner part of the solar system where we can see them. At this point they are literally being slung about the sun by gravity and then back out into the deepfreeze where they were first formed some 4 to 5 billion years ago.

The possibility of a comet's striking the earth, as opposed to a meteor, is considered extremely small, but not impossible, and in fact, the catastrophism theory of Immanuel Velikovsky employs such a mechanism.

The other theory of cometary birth is far more violent and had its genesis in the appearance in 1969 of a different type of comet. It looked at first blush like your typical hairy star (*kometes* is Greek for "long-haired") with its flaming tail streaming behind, but its brightness and what could be deduced of its structure were very different. The coma—the bright, spherical halo a comet wears about its head—was exceptionally large and seemed to be increasing. One of the world's expert comet watchers, Czech astronomer A. Mrkos of the Charles University in Prague, decided that this was no ordinary comet. "The comet," he told the Twenty-first Nobel Symposium in Stockholm, "was captured by Jupiter in a not too distant past. Consequently the comet is very young and at the stage of formation in a new orbit around the sun."

The problem with Mrkos' deduction, of course, is that comets are supposed to be very old, indeed as old as the solar system and composed of its most primitive materials. But Mrkos' idea that this comet—dubbed 1969e, it being the fifth comet observed in that year—is young has some theoretical support in the ideas of a Russian astrophysicist, Sergei Vsekhsvyatskii. He believes that comets are constantly being born in volcanic eruptions on the planets. The explosions hurl the comets into the solar system, where they are seized by either the sun's gravity or Jupiter's and pulled on their journeys around the planets.

There are certainly enough active volcanoes sited about the solar system to provide the many wombs that would be required to put into space the 2000 or so comets that have been observed over the last 300 years. The probes of Jupiter, Mars, Venus and Mercury indicate at least as much volcanic activity as on the earth. The question is: Are these volcanoes powerful enough to hoist a chunk of gas and matter the size of a comet—and some are quite large—into space? To overcome Jupiter's massive gravity, for example, a formidable escape velocity of 60 km per second would have to be achieved. On earth, however, escape velocity is only 11.2 km per second, while on the smaller planets, such as Mars and the moon, escape velocity is even less. Vsekhsvyatskii has calculated that tremendously powerful volcanic explosions such as Krakatoa, which erupted in 1883 and darkened half the earth for six months after with its fallout, generated $10^{28}$ ergs—enough energy, he says, to place a very respectable-sized comet into orbit. But according to the Soviet scientist, the earth and the other inner planets of the solar system are too cold to generate the energy necessary to keep the solar system stocked with an abundant supply of comets. For that, the outer gas giants are needed, and in Vsekhsvyatskii's eyes they are the primary sources of the comets that fly about our solar system. In Vsekhsvyatskii's theory the planets were not coalesced from clouds of dust and gas but were instead thrown off from the sun. In this way, they all were originally made of a solar type of material and had at the time of creation roughly equal masses and rotational energies. By subtracting these masses from the planetary masses observed today, Vsekhsvyatskii arrives at a number very close to his estimate of the total mass of past and present comets and meteors in the solar system. "This," he insists, "makes it quite certain that the planets could not have originated from cold bodies condensed from a gas-dust medium. They must rather have been bodies of a stellar nature."

And so, by the medium of catastrophic ejection, Vsekhsvyatskii accounts not only for the birth of comets, but for the creation of the solar system.

While the theory must still be proved, and may well be by future space probes to the comets themselves, these remarkable bodies have provided some of the more extraordinary spectaculars the heavens offer. And while the ancients cowered in fear and saw evil omens in each comet as it flashed across the night sky, one man saw a certain repetition in their passage. Edmund Halley, a student of Isaac Newton's, thought that comets traveled like the planets in predictable orbits about the sun. Using Newton's formulas, he calculated the paths of several comets and found that three—reported in 1456, 1531 and 1607—had approximately the same orbit as one he himself had observed in 1682. Halley believed they were one and the same comet, and he boldly predicted it would appear again in seventy-six years, the time it took for its complete passage about the sun. In 1758 Halley's Comet showed up, right on schedule, and another victory for predictability and order in the universe had been gained.

Now another man has been calculating the periodic orbits of Halley's Comet and found a discrepancy between his computer calculations and the reported observations of past orbits. Joseph Brady is an astronomer at the University of California's Lawrence Radiation Laboratory; in 1971, after tracing each orbit of Halley's Comet since A.D. 295, he found discrepancies of two to three months in its appearance. The reason, Brady concluded, was that Halley's Comet was being slowed in its orbit by the gravitational attraction of another, unknown and hitherto-unsuspected planet. He dubbed it Planet X and determined that it was located in an orbit beyond Pluto some 55 billion miles from the sun and that it had a mass of about three times that of Saturn. Gravitational perturbations in the orbits of known bodies have been used in the past to predict unseen planets. The location of Neptune was predicted in 1846 on the basis of perturbations in the orbit of Uranus, then the farthest of the known planets. Then, in 1915, Neptune's orbit was found to be perturbed, and Pluto was finally sighted in 1930.

And what of Planet X? "The proposed planet is located in the densely populated Milky Way," says Brady, "where even a tiny area encompasses thousands of stars. If it exists, it will be difficult to find."

Indeed. During June and July 1972 astronomers at England's Royal Greenwich Observatory conducted a photographic search for Planet X using a 13-inch telescope. The photographs covered an area extending three and a half degrees from the predicted position in every direction. The results: nothing. "If a trans-Plutonian planet does exist," reported Greenwich astronomer A. P. O. Foss, "then either it is a much less massive object, and hence considerably fainter than Brady has predicted, or it is not near Brady's final position."

This is not the first time planets have been predicted

and not found, nor are gravitational perturbations the only means of predicting their existence. In 1772 a young German astronomer named Johann Elert Bode published an astronomy text, *Introduction to the Study of the Starry Sky,* in which he took note of a remarkable mathematical coincidence:

"Let the distances from the Sun to Saturn be taken as 100 then Mercury is separated by 4 such parts from the sun. Venus is 4 + 3 = 7; the Earth is 4 + 6 = 10; Mars 4 + 12 = 16. Now comes a gap in this progression. After Mars there follows a distance of 4 + 24 = 28 parts in which no planet has been seen. . . . From here we come to the distance of Jupiter at 4 + 48 = 52 parts and finally to that of Saturn at 4 + 96 = 100 parts."

Set down on a table, the Titius-Bode law (Johann Daniell Titius, a professor of mathematics in Wittenberg, Germany, had discovered the same planetary relationship at about the same time) shows some remarkable similarities to the actual distances of the planets from the sun. In the case of this table, we shall measure distances in astronomical units (AU) which are defined as the mean distance, 93 million miles, between the earth and the sun. The Titius-Bode numbers are expressed in tenths.

| | TITIUS-BODE LAW | ACTUAL AU DISTANCE |
|---|---|---|
| MERCURY | 4 + 0 = .4 | 0.39 |
| VENUS | 4 + 3 = .7 | 0.72 |
| EARTH | 4 + 6 = 1.0 | 1.00 |
| MARS | 4 + 12 = 1.6 | 1.52 |
| MISSING PLANET | 4 + 24 = 2.8 | —— |
| JUPITER | 4 + 48 = 5.2 | 5.20 |
| SATURN | 4 + 96 = 10.0 | 9.54 |
| URANUS | 4 + 192 = 19.6 | 19.18 |
| NEPTUNE | 4 + 384 = 38.8 | 30.06 |
| PLUTO | 4 + 768 = 77.2 | 39.44 |

When the Titius-Bode law was formulated, Uranus was the farthest known planet, and with the exception of the gap between Jupiter and Mars, all the planets fitted the law. But Bode's law begins to unravel at that point. Neptune is somewhat closer, 8 AU to be exact, than Titius-

Bode predicts. The error of Neptune could be fudged somewhat, but not that of Pluto. At 4 billion plus miles from the sun, almost 40 astronomical units, Pluto is badly out of position if the law or relationship is to be considered reasonably valid. But if Pluto is not properly a planet, but merely an escaped moon of Neptune, as most astronomers believe, then we must look farther out for a planet, if indeed there is one. According to Bode's law, the ninth planet in our solar system (we have already ruled out Pluto as a planet) must be not at 39.44 AU from the sun, where Pluto resides, but almost twice as far away, at 77.2 AU. And this, in fact, is where astronomer James Brady's computer calculations insist Planet X will be found, some 5 billion miles beyond Pluto, or roughly 78 AU from the sun.

If Bode's law is correct, does that mean that order is restored and catastrophe abjured by nature within our solar system? Not if we attempt to find that missing planet between Mars and Jupiter that Bode's mathematics insists should be there. That search was actually begun soon after Bode's law was published in 1772. A convention of astronomers meeting in 1796 decided to search the region between Mars and Jupiter for evidence of the missing planet. Four years later an astronomer who did not attend the conference, Giuseppe Piazzi, director of an observatory on the island of Sicily, was attempting to correct a star catalogue. On the first night of the nineteenth century he found a star not listed in the catalogue and one which he had never seen before. The next night he saw it again, but in a slightly different position, which meant it had moved, meaning it could not be a star. Piazzi thought it was a comet, but Johann Bode convinced him that it was a planet, in fact, *the* missing planet the law predicted would lie between Mars and Jupiter.

Piazzi was willing to be convinced; after all, the discovery of a new planet is the sort of thing that gets one listed in history books, while the discoverers of comets—except for Edmund Halley—are usually mere footnotes in astronomy catalogues. Unfortunately Piazzi fell ill before he could complete his observations of the new planet, and when he recovered, it was gone. He had been unable to gain enough data to work out its orbit, and so no one was able to find his new planet. Then a brilliant German mathematician, Karl Friedrich Gauss, figured out the orbit of the new planet. At the time, however, it was too close to the sun to be seen. Then, almost a year to the day Piazzi had first sighted his planet, it was found again, following the orbit Gauss had predicted.

Poor Piazzi, it was named Ceres, after the Babylonian goddess of the harvest, not for the Sicilian astronomer. But there was more bad news, for Ceres turned out to be so diminutive, a mere 478 miles in diameter, that it could scarcely be classed as a planet and was instead declared an asteroid, a minor planet, a mountainous rock tumbling in an orbit about the sun. On the heels of the Piazzi discovery, astronomers swiftly found three more asteroids very close to Ceres. Eventually more than 1600 asteroids were discovered moving in orbit about the sun and filling the space between Mars and Jupiter, exactly where the Titius-Bode law predicted a planet should be. All told, astronomers "guesstimate" there may be as many as 50,000 asteroids in the belt between Mars and Jupiter. Is this vast asteroid belt the remains of a planet? Are the flying mountains, rocks, pebbles and dust of the belt the evidence of an incredible catastrophe that literally blew a planet to pieces? If so, it would mean that virtually on our doorstep, just one planet removed from earth, there lies incontrovertible proof that planets are subject to destructive forces easily powerful enough to tear the earth itself to bits.

But while astronomers and others have drawn some rather fanciful scenarios to account for the destruction of what I shall call Bode's Planet, there is little in the way of hard evidence. The hardest is, in fact, in the form of bits and pieces of the asteroid belt that are pulled to earth by its gravitational attraction. Many of the meteorites that strike the earth are believed to be chunks of debris from the asteroid belt, and they come for the most part in two types—metallic and stony. This, of course, is precisely what we would expect to find in the debris of a planet like the earth with its metallic core and stony mantle covering the core.

This in turn has given rise to the intriguing idea that a highly intelligent form of life developed on Bode's Planet and blew its world apart in a nuclear holocaust. Needless to say, that idea is a post-Hiroshima warning, but the force needed to blow apart a planet, even a small one, is so great that it would probably vaporize it completely, leaving no debris at all.

Other theories suggest a collision with another planet, but here again the destructive forces involved would be so vast as to cause total vaporization of the planet. The final and supposedly fatal blow to the exploding planet theory is the fact that the mass of all the 50,000-plus asteroids and all the meteorites that have struck the earth does not equal a planet of even modest dimensions. It is this argument that is the main support of the stillborn planet school of thought, which says that the asteroid belt is composed of planetary materials that failed to accrete into a full-scale planet. "For some reason," points out Dr. Robert Jastrow, director of the Goddard Institute for Space Studies, "these planetesimals did not reach the ultimate stage of accumulation into a planetary body, as did the other objects in the solar system. . . ." Still, there remains the nagging doubt that all those asteroids might once have been a planet, for Jastrow goes on to note, "or if they did, they were disintegrated again in a subsequent catastrophe."

Such a catastrophe could well have been authored by Jupiter, which, as we shall see in a subsequent chapter, is blamed for one of the earth's great catastrophes. Jupiter, as the most massive body in the solar system save for the sun, acts as a giant gravitational vacuum cleaner. Some of its moons are thought to have been captured from the asteroid belt some billions of years ago. But Jupiter's great mass, converted into gravitational pull, makes it powerful enough literally to tear a protoplanet to pieces before it can build the necessary mechanical strength to withstand such brute force. Then the gravitational vacuum cleaner of Jupiter might have swept up much of the debris that littered the space nearby. Mountains, even planetary pieces the size of the moon, could fall into the enormous maw of Jupiter without a trace. Indeed, so vast is it that Jupiter could swallow a dozen earths without a burp. This enormous mass makes Jupiter a menacing planet that has figured in the fears of people since their first attempts to understand the universe. The mythic stature of Jupiter looms large in almost every culture whose written or oral traditions have come down to us. In some cases these mythologies reach back into prehistory and warn of an inchoate, unmentionable menace in Jupiter.

How valid is that menace? Just recently we have taken the closest look we have ever had of the stillborn star called Jupiter.

# 3

# WORLDS IN COLLISION

On Monday, December 3, 1973, man reached out and virtually touched the most awesome of the gods in his classical pantheon—the planet Jupiter. For on that day a tiny, 570-pound spacecraft survived a violent storm of radiation and passed within 81,000 miles of the largest planet within our solar system.

A few weeks later another robot explorer, Mariner 10, passed by Venus and looped around Mercury before heading off into orbit about the sun. The information both spacecraft sent back to earth amazed and puzzled the scientists, who began immediately to pore over the data with unrestrained delight. And from the millions of informational bits returned from the close-up study of the three planets and the startling candid photos of their faces there begins to emerge a mountain of detail that both confirms and contradicts many of the ideas we once held about Jupiter, Venus and Mercury. At the same time the information, added to all our other knowledge collected in the space age, provides still more evidence to support the controversial theories of a man who has questioned the immutability of the solar system and challenged such gods of science as Newton and Darwin with such statements as:

"Several times during the 15th and 8th centuries B.C., the

earth was convulsed by near collisions with other celestial bodies. These cosmic brushes caused a series of catastrophes that altered the course of ancient history.

"A host of ancient myths and legends offer clues as to what happened during these catastrophes. Further written evidence is contained in the Old Testament Book of Exodus and also historical and astronomical texts inscribed on papyrus and on stone and clay tablets.

"The planet Venus originated in a violent disruption of Jupiter. Venus must be exceedingly hot.

"The universe is not a vacuum populated only by celestial bodies. Rather, it is crisscrossed by charged particles and riven by magnetic fields."

These and other ideas were first published in 1950 in a book called *Worlds in Collision.* Its author was Immanuel Velikovsky, a medical doctor and psychiatrist, then little known outside professional circles. His ideas were considered nonsense by most historians and astronomers of 1950 and stirred a storm of controversy that has not died to this day. So violent was the initial reaction to the Velikovsky book that it went beyond mere rejection of an idea and bordered on intellectual persecution, reminding some scholars of the trials of Bruno and Galileo, who also held ideas that did not agree with accepted dogma.

Immanuel Velikovsky is a tall, gaunt white-haired scholar of immense persistence. Born in Russia in June 1895, he attended school in Moscow. Barred from entering Moscow's Imperial University by clauses restricting Jews, he began premedical studies in Scotland in the spring of 1914. At the outbreak of World War I he was on summer vacation in Moscow. Unable to return to Scotland, he studied ancient history and other humanistic fields at the Free University, now known as the Karl Marx Institute. In 1916 he was finally admitted to Moscow University, which in 1921 awarded him a medical degree. He moved to Berlin, where he founded and edited *Scripta Universitatis,* a series of monographs by outstanding Jewish scholars the world over; Albert Einstein was a notable contributor. By 1924 Velikovsky had moved to Jerusalem, where he began the practice of medicine. Later he studied psychoanalysis in Vienna under Sigmund Freud's first pupil, Wilhelm Stekel.

Velikovsky's scholarly mind also led him to write a number of papers for Freud's psychology journal, *Imago.*

Among them was Velikovsky's then-startling idea that distortions in electrical brain wave patterns would be found in epileptic patients. Velikovsky had met Freud, and the two men exchanged letters for many years. Then Freud published a series of essays called *Moses and Monotheism* in which he sought the psychologic wellsprings of religion. Velikovsky began a complementary study of the three subjects of Freud's paper, Oedipus, Akhnaton and Moses. Entitled *Freud and His Heroes,* the book was also to have contained an analysis of Freud's own dreams.

To research the book, Velikovsky took his family to New York in the summer of 1939. While doing this research, he came to the realization that the biblical story of Moses was not merely allegory or simple myth but could perhaps be an account of events that had actually taken place. The more he pursued the idea, the more evidence he amassed to indicate that a great catastrophe had taken place, one that had affected not only Moses and the Jews, but the entire world. And if that were the case, there should, he reasoned, be other accounts, especially an Egyptian history that would match that of the biblical tale of Exodus. Consumed by the idea, Velikovsky began searching for an Egyptian account of the Exodus. Finally, after many weeks he discovered a translation of an Egyptian papyrus which contained not only a description of a great catastrophe, but a precise accounting of the plagues of Egypt.

The account was called *Admonitions of a Sage,* or the Ipuwer papyrus. Dr. Velikovsky offered the following examples of agreement between the two ancient texts:

EXODUS 7:21. . . . there was blood throughout all the land of Egypt.

PAPYRUS 2:5–6. Plague is throughout the land. Blood is everywhere.

EXODUS 7:24. And all the Egyptians digged round about the river for water to drink; for they could not drink of the water of the river.

PAPYRUS 2:10. Men shrink from tasting—human beings thirst after water.

EXODUS 7:21. . . . and the river stank.

PAPYRUS 3:10–13. That is our water! That is our happiness! What shall we do in respect thereof? All is ruin!

EXODUS 9:25. . . . and the hail smote every herb of the field, and brake every tree of the field.

PAPYRUS 4:14. Trees are destroyed.

PAPYRUS 6:1. No fruit nor herbs are found. . . .

EXODUS 10:15. . . . there remained not any green thing in the trees, or in the herbs of the fields, through all the land of Egypt.

PAPYRUS 6:3. Forsooth grain has perished on every side.

PAPYRUS 5:12. Forsooth, that has perished which yesterday was seen. The land is left over to its weariness like the cutting of flax.

EXODUS 9:3. . . . the hand of the Lord is upon thy cattle which is in the field . . . there shall be a very grievous murrain.

PAPYRUS 5:5. All animals, their hearts weep. Cattle moan.

EXODUS 9:19. . . . gather thy cattle and all that thou hast in the field.

EXODUS 9:21. And he that regarded not the word of the Lord left his servants and his cattle in the field.

PAPYRUS 9:2–3. Behold, cattle are left to stray and there is none to gather them together. Each man fetches for himself those that are branded with his name.

EXODUS 10:22. . . . and there was a thick darkness in all the land of Egypt. . . .

PAPYRUS 9:11. The land is not light . . .

EXODUS 12:29. And it came to pass, that at midnight the Lord smote all the firstborn in the land of Egypt, from the firstborn of Pharaoh that sat on his throne unto the firstborn of the captive that was in the dungeon. . . .

PAPYRUS 5:3, 5:6. Forsooth the children of princes are dashed against the walls.

PAPYRUS 6:15. Forsooth, the children of princes are cast out in the streets.

EXODUS 12:30. . . . there was not a house where there was not one dead.

PAPYRUS 2:13. He who places his brother in the ground is everywhere.

EXODUS 12:30. . . . there was a great cry in Egypt. . . .

PAPYRUS 3:14. It is groaning that is throughout the land, mingled with lamentations.

With this discovery Velikovsky began looking for other parallels not only in Egyptian history, but throughout the world, and as he found them, he realized that a great natural disaster had altered the course of ancient history, a disaster that had never been recorded as a historical event. But why not? Surely a disaster of such magnitude would be indelibly seared in man's history. To answer this question, Velikovsky looked to his psychoanalytic training. He knew of many cases where individuals blocked from their minds the memories of extremely painful experiences. So, too, he reasoned, might an entire race blot out the recollection of a devastating catastrophe that virtually destroyed their society and most of its people. He called such a process collective amnesia and began a monumental life's work, a reconstruction of ancient history according to his catastrophe theory.

Settling with his family in Princeton, New Jersey, in 1939, he continued the massive research that led to the initial publication of *Worlds in Collision.* Now eighty, Dr. Velikovsky has been buffeted, like a towering oak, by storms of unbelievable fury and savagery as the entire scientific community has mounted assault upon attack, ridicule upon scorn and vituperation to discredit his theories. For Velikovsky, citing historical and paleontological evidence, was attacking the very rules by which such towering scientific figures as Galileo, Copernicus, Newton and Einstein had said our solar system was governed. And yet, by following a scholarly trail of myth and legend across the world and fitting it to astronomical events, Velikovsky was destroying the immutability of the dogma that decreed the unchanging, unceasing orbits of all the planets of our solar system around its sun since time began. And by bridging the enormous gaps between such soft disciplines as history and literature and such hard sciences as physics and astronomy, Velikovsky was compounding his heresies in the eyes of the scholars and scientists of both worlds.

"One of the reasons Velikovsky elicits the kind of reaction he does from other scientists is his method, which is somewhat different from the ordinary methods of empirical research," explains Dr. Lionel Rubinov, professor of philosophy at Trent University in Canada. "He starts with myth and literature, developing hypotheses from these areas which he then applies to the interpretation of natural phenomena. His approach has been to speculate rather than to perform experiments. The incredible thing is that when experimental data finally is produced, it tends to confirm his hypotheses."

It is precisely this use of myth and literature as the starting point for physical theory that has so enraged scientists over the past quarter of a century. For it is not a simple, straight-line relationship Velikovsky seeks to establish, but a complex, interdisciplinary thesis that ranges from the chemistry of the Martian atmosphere to the genesis of the "plumed serpent" of Mexican mythology, from the chemical formula of manna to the geological events that brought the most recent ice age to a close, from the origin of species to the identity of the Queen of Sheba, and so on and on from the deepest recesses of man's inner torment to the outer reaches of our solar system.

Velikovsky's theories bridge such widely diverse disciplines as physics, mythology, genetics, psychology, archaeology, astronomy, paleontology, history, geology and chemistry in seeking to explain and understand the history of our world in the light of the catastrophes it has encountered.

From the memories stored in myth and legend, from the testimony of broken pillars and archaic calendars, from the records of bones and stones, Velikovsky concluded that the vision of a safe, orderly, unchanging universe, with planets and satellites moving in clockwork precision since the beginning of time, was wrong. And if the earth's path about the sun had been changed or interrupted, it was logical for Velikovsky to speculate on both the causes for the change and the effects they would have on the planets and on the creatures that lived upon them.

Everything since—his reconstruction of history, his collection of geologic evidence in another of his books, *Earth in Upheaval,* to prove further that the earth had indeed suffered great catastrophes in its past, his challenges to Darwinian evolution and his application of psychology to national behavior—is the product of a mind that recognizes no boundaries on where its conclusions are to lead. And even though the Velikovsky theories are still largely scorned by the scientific establishment, the interdisciplinary approach to understanding the physical universe in which our earth moves and we live is increasingly being taken by scientists eager for answers that can be obtained only by pushing at the borders of their own specialties. The tools of chemistry find increasing application by the geologist and biologist, for example. The biologist has become a biochemist, a biophysicist, a molecular biologist and a psychobiologist. While these fields are often considered new disciplines sacrosanct unto themselves, they are, in fact, new bridges between older disciplines.

The truly universal application of Velikovsky's theory can perhaps best be seen in the wide-ranging number of doctoral theses now being written in universities in an equally wide-ranging number of faculties—ancient history, sociology and psychology, the history and philosophy of science, the history of religion, geology, biology, the theory of evolution, cosmology, planetary sciences and celestial mechanics. When Neil Armstrong first set foot upon the moon in 1969, it was a triumph of engineering technology, but the rocks he brought back from the lunar surface represented a history of the moon since its beginnings and clues to the origin of the earth and solar system that are still being explored by scientists around the world.

In those rocks, bearing the imprint of particles of the sun borne by a solar wind perhaps billions of years ago or perhaps within moments of their being lifted from place, and in the rocks of the earth are links that may have been forged by the same event. It is Dr. Velikovsky's contention that these links can be seen and uncovered only by utilizing all the tools of man's wisdom, not just a few of his individual disciplines.

Initially, the ideas contained in *Worlds in Collision* found critics even among the most strident combatants of the then hotly raging "cold war." The New York *Daily News* called it a Russian propaganda plot; the communist newspaper then published in New York, the *Daily Worker,* saw acceptance of the theory as evidence that the bourgeois society of America had entered its death throes. Even so towering an intellect as the British biologist J. B. S. Haldane considered the book an attempt by warmongering American munitions makers to prepare the world for the nuclear holocaust they were planning to launch.

Yet it remained for the scientific community to launch the most vicious and unreasoning attack on both the ideas and the author of *Worlds in Collision.* Chief among the attackers was Dr. Harlow Shapley, a world-renowned astronomer who was then director of the Harvard Observatory. Shapley had been introduced to Velikovsky in 1946 and declined to read his manuscript, pleading the pressure of other affairs. If, however, certain other scientists were to read and approve it, Shapley said, he would then perform some spectroscopic analysis Velikovsky had suggested that might help prove his theory.

"Among the tasters mentioned to protect Shapley from intellectual poisoning," recalls Dr. Horace Kallen, an eminent scholar and co-founder of the New School for Social Research, "I was one, and Shapley agreed that if I read Velikovsky's manuscript first and recommended it, he too would read it."

Kallen read the manuscript and was so impressed he urged Shapley to perform the spectroscopic analysis which would either further the theory or disprove it.

Shapley declined and thought that was the end of it. But when, in 1949, Macmillan announced it was about to publish *Worlds in Collision,* Shapley sought to block its release.

In a subtly worded letter to Macmillan he threatened to "cut off" relations with the publisher. In what appeared to be an organized boycott, Shapley's threat was followed by letters from other scientist-authors and professors who used Macmillan's textbooks for their courses. By then, however, *Worlds in Collision* was already on the presses. The head of Macmillan thanked Shapley for his concern and offered to have the book rechecked by new scientific readers. Two of the three advised publication, and the book was released.

In the interim the manuscript had been circulated to several periodicals, including *Harper's* magazine and the *Reader's Digest.* Commenting editorially on an article on the book by Eric Larrabee, *Harper's* noted, "No one who has read Mr. Larrabee's article can ever again read the Old Testament prophets with the same blind piety or same blind skepticism that he felt before."

But blind, unreasoning disbelief was all the scientific community accorded the Velikovsky thesis, blind to the point of refusing even to read the book. "No, I have not read the book," admitted Dean B. McLaughlin, professor of astronomy at the University of Michigan, in a letter of protest to Macmillan. "I have read the ballyhoo that preceded it, written by such 'authorities' as Eric Larrabee and Fulton Oursler [author of the *RD* article], whose ignorance is exceeded only by their colossal conceit."

While some scientists were attacking not only the Velikovsky thesis, but all who supported it, others stood firm in their support. Dr. Gordon A. Atwater, chairman and curator of the Hayden Planetarium of the American Museum of Natural History, commented in a preface to the Oursler article in the *Reader's Digest* that in the light of the Velikovsky thesis "the underpinning of modern science can now be re-examined."

Atwater was so intrigued by the idea that he planned to mount a star show at the planetarium illustrating the *Worlds in Collision* theory. He also wrote a review of the book in which, although skeptical of some of Velikovsky's findings, he took the records of worldwide catastrophes in historical times to be evidential and urged that the book be read with an open mind. The night before the review was published, and before he could produce the planetarium show, he was dismissed from both his positions with the museum.

Soon after *Worlds in Collision* was published, the editor who had bought the book for Macmillan and who had been with the company for twenty-five years was fired. But the book was published and quickly became a best-seller.

The reviews by scientists, in journals and the popular press, were notable for their savagery and denunciations, in direct contrast with the spirit of scientific inquiry, for none of the reviewers had read the book. Nor were they content simply to denounce the book. They stepped up the pressure on the publisher to abandon *Worlds in Collision*. Macmillan salesmen began to get the cold shoulder from science professors and teachers in universities all over the country. Macmillan got the message and eight weeks after publishing *Worlds in Collision* transferred publication rights to Doubleday, a publisher without a textbook division. The sale was unparalleled in publishing history, for *Worlds in Collision* was then "Number One" on the New York *Times* best-seller list.

It was small solace to Immanuel Velikovsky, however, because the attacks continued and no "reputable" scientific journal would accept a paper from him in reply. But the scientific establishment that was so swift in its denunciations was also making possible the means by which many of the Velikovskian predictions about the state of the planetary bodies could be tested. With the birth of the space program men and machines were sent to those very worlds Velikovsky insisted had collided with earth.

And even before we began sending our missiles into space, evidence was pouring in on all who listen. "Space," Velikovsky had declared, "is not a vacuum; and electromagnetism plays a fundamental role in our solar system and the entire universe." Although some stars were known to give off radio waves, the idea of a noisy space, crackling with radio waves pressed by magnetic fields and riven by electrical charges and radioactivity, was not a widely accepted part of the astronomy of 1950. Thus, few astronomers gave any credence to Velikovsky's claim in a 1953 lecture at Princeton University that Jupiter was emitting radio noises.

Albert Einstein was sympathetic to some of Velikovsky's fundamental concepts but vigorously opposed his theory that space was permeated by magnetic fields, that the sun and planets are charged bodies and that electromagnetism plays a role in celestial mechanics.

In June 1954 Velikovsky offered in writing to stake the outcome of his debate with Einstein on the question of whether Jupiter emits radio noises, as he had claimed. Einstein replied, as was his custom, by making marginal notes, one of which discounted the idea.

Ten months later, early in 1955, astronomers at the Carnegie Institution were shocked to hear strong radio signals pouring in from Jupiter. When Einstein heard the news, he emphatically declared that he would use his influence to have Velikovsky's theory put to experimental test. Nine days later he died—a copy of *Worlds in Collision* open on his desk.

Our picture of Jupiter has been vastly expanded since then, although huge rents remain in our knowledge of that far-off world. But the twenty-five years of astronomical advances and mountains of information just sent back by Pioneer 10 are helping fill in the gaps. Astronomers have long known that the mass of Jupiter is sufficient to provide enough material for not one but dozens of planets the size of Venus, for it is the most massive of the planets, containing 71 percent of all the planetary matter in the solar system, and is exceeded in size only by the sun itself.

Velikovsky referred to Jupiter as a "dark star," and astronomers today for the most part agree that Jupiter is probably a "stillborn sun, a body that fell just short of igniting a thermonuclear furnace." The voyage of Pioneer 10 offers further confirmation of this, for it measured huge, periodic bursts of highly charged particles pouring from Jupiter. Moreover, Pioneer 10 began to pick up these signals while still six months away from its target as Jupiter scrawled its signature across 100 million miles of sky.

"It looks," said the University of Chicago's John Simpson, who had been analyzing the signals, "as though we have gone from a region where the sun dominated to a region where Jupiter dominates."

It is a planet of awesome energies radiating almost three times more energy than it gets from the sun. Pioneer 10 measured an enormously powerful magnetic field about the planet which held within its force fields radiation belts of enormous energies—fluxes of high-energy electrons in such vast number that they exceeded by a thousand times the most optimistic of earth-based predictions.

Data such as these delighted the astronomers and planetologists who look on Jupiter as a sort of Rosetta stone of the solar system and on Pioneer 10 as the source of astonishing surprises. Yet none of the newly discovered information about Jupiter surprises Velikovsky, for he believes it confirms his predictions and neatly fits the demands of his theory.

Not the least of these predictions was that Jupiter's core would have a high temperature. This comment was made at a time when astronomy textbooks were asserting that Jupiter was an icy planet with a surface temperature about –150 degrees Fahrenheit.

Pioneer's temperature readings of –207 degrees F recorded at Jupiter's cloud tops seemed to bear that out, for the temperatures would tend to drop even more at the upper reaches of the atmosphere, where whatever radiant energies were stored at the core would be rapidly dissipated as they percolated up through the 25,000-mile-deep atmosphere. But as the spacecraft probed deeper, it recorded a temperature of 260 degrees F, or 48 degrees above the boiling point of water, only 125 miles down into the

Jovian atmosphere. Not much farther down, on the night side, where no sunlight could reach, the temperature soared to 800 degrees. This discrepancy astonished the scientists analyzing the data, and some estimated that at its core Jupiter might have a temperature of 2500 degrees F, or even as much as 10,000 degrees F.

As Pioneer 10 passed by Jupiter, it clocked the turbulent, seething clouds of its atmosphere whirling around the planet at an incredible 22,000 miles per hour. Here then, according to Velikovsky, on a massive, seething planet where turbulent clouds trigger lightning discharges equal to the power of two exploding hydrogen bombs, where radiation fields 10 million times stronger than those girdling the earth fling unimaginable energy forces about, Jupiter, the emperor of the planets and the chief of ancient man's deities, underwent a shattering convulsion and hurled a planet-sized chunk of itself into space. According to Velikovsky's interpretation of ancient texts, a blazing comet, the protoplanet Venus, was torn from Jupiter 4000 years ago and hurtled down a long orbit toward the sun, an orbit that sent it on a course that would menace the earth.

To Immanuel Velikovsky this fiery birth of Venus was recorded by terrified groups of people all over the earth. "In every country of the ancient world we can trace cosmological myths of the birth of the planet Venus," he wrote in *Worlds in Collision*. "In Greece the goddess who suddenly appeared in the sky was Pallas-Athene. She sprang from the head of Zeus-Jupiter."

The violent paroxysm of birth that tore Venus from Jupiter would require such fantastic energies that some scientists have considered the Velikovsky theory impossible on that issue alone. Others point to a still-popular astronomical theory that virtually all comets originate outside the solar system in a vast swarm of cosmic debris that stretches beyond Pluto halfway to the nearest star, Proxima Centauri. But for the past forty years a famous Russian astronomer, Dr. Sergei Vsekhsvyatskii, director of the Kiev University Observatory, has attempted to prove that the planets of our solar system have themselves given birth to as many as 100 billion comets in the last 10 million years. (See Chapter Two.)

He argues that enormous energies are generated as the planets eject pieces of themselves into space. As proof of cometary birth on the planets of the solar system, Dr. Vsekhsvyatskii offers "the fact that the presence of ammonia-methane-ice in comet nuclei has now been demonstrated is a direct indication of eruptive processes on the planets." The same ice mixture has often been found on the surface of the earth, Mars and some satellites of Jupiter, which lead to only one conclusion, according to Dr. Vsekhsvyatskii: "The comets must be eruption ejecta from the planets."

The site on Jupiter where such an incredible event might have occurred is marked by a huge red spot, photographed by Pioneer's cameras. The Great Red Spot, which is also visible from earth telescopes, measures 15,000 miles across and 8,000 miles high, an area large enough to embrace earth and Venus and four other planets of similar size.

Some millennia after being torn from Jupiter the protoplanet Venus was thrown into an elliptical orbit around the sun. Along that orbit, some 480 million miles from its birthplace on Jupiter, it would intersect the orbit of the earth. As the flaming daughter of Jupiter headed toward the sun, and inexorably the earth as well, men watched in horror while the heavens seemed to catch fire. To the Babylonians who witnessed the birth of this new goddess, it was the "bright torch of heaven." To the Chinese, Venus spanned the heavens, rivaling the sun in brightness. "The brilliant light of Venus," noted one ancient rabbinical record, "blazes from one end of the cosmos to the other."

In the middle of the fifteenth century B.C., Velikovsky theorized, earth in its orbit about the sun entered the outer edges of the protoplanet's tail of dust and gases. A fine red dust filled the air, staining the continents and the seas with a bloody hue. The rusty dust particles chafed and rubbed away human skin; people grew ill and fell as they walked. Fresh water in brook, stream and river turned red and foul-smelling as fish died and rotted by the millions. Frantically men clawed at the earth, seeking underground streams uncontaminated by the red dust.

"And all the Egyptians digged round about the river for water to drink; for they could not drink of the water of the river," says Exodus 7:24.

"The river is blood. . . . Men shrink from tasting—human beings thirst after water," confirms the Egyptian sage

Ipuwer. "That is our water! That is our happiness! What shall we do in respect thereof? All is ruin!"

The earth continued to move through the cometary tail and the rain of dust grew heavier, the particles coarser and larger until it seemed the world was bombarded by whirlwinds of gravel and stones. Exodus recorded the change in the character of the dust as yet another disaster falling upon the Egyptians. "So there was hail, and fire mingled with the hail, very grievous, such as there was none like it in all the land of Egypt since it became a nation." (9:24)

The hail of stones sent men fleeing in panic, abandoning their fields and flocks as they sought cover from the lethal bombardment. Crops were smashed flat, cattle dropped as if poleaxed, "and the hail smote every herb of the field, and brake every tree of the field," according to the account in Exodus.

Ipuwer concurs: "Trees are destroyed . . . no fruit nor herbs are found. Forsooth, that has perished which yesterday was seen. The land is left to its weariness like the cutting of flax."

These things happened, say the Mexican Annals of Cuauhtitlán, indicating a global rather than a local disaster, when the sky "rained not water, but fire and red hot stones."

The shower of micrometeoroids continued to pelt the earth as it moved through the comet's tail, but now an even more terrifying event took place. The sky seemed to burst into flame as hydrocarbon gases, a primary constituent of the inner portions of the tail, according to Velikovsky, ignited in fiery gouts that rent the sky. In some parts of the earth unignited petroleum streamed down, sinking into the land and floating on the waters. Vast lakes of naphtha spilled down upon Siberia, the Caucasus and the sands of Asia Minor, shrouding the region in smoky fires that provided the dark backdrop to a suffering humanity.

The smoking, gloom-shrouded earth continued its careering orbit through the comet's tail on a collision course with the massive head.

Suddenly, says Velikovsky, a violent convulsion ripped the earth; it tilted on its axis in the grip of the protoplanet's gravitational pull; the terrestrial crust pulled and shifted. Cities were leveled, towering forests snapped like toothpicks, mountains melted, islands shattered, oceans crashed over continents, and most of the earth's animal and human populations were destroyed.

"The lightnings flashed and the thunders roared terribly and all were afraid," states the tradition of the Cashinaua of western Brazil. "Then the heavens burst and fragments fell down and killed everything and everybody. Heaven and earth changed places. Nothing that had life was left upon the earth."

Rivers reversed their direction, and the shattered earth was pummeled by a global hurricane which flattened whatever was still standing. In China the waters "overtopped the great heights, threatening the heavens with their floods." In the Aegean a mighty island exploded, sending pumice and ash raining down over the shores of Africa. Continental plates shifted, mountains moved, desert replaced farmland as the earth convulsed and writhed under the blows of the huge comet's gravitational pull.

With each convulsion, Velikovsky believes, the axis of the earth tilted until suddenly an equilibrium was reached between the comet's pull and the earth's rotational velocity. It was as if the earth were suddenly frozen in space, locked in place in its orbit like a stone lodged in a tire tread, still going around and around, but having no movement of its own. Held in a gravitational thrall, its axis tilted, part of the world hung in extended darkness, part in protracted day.

The Persians watched in awe as a single day became three before turning into a night that lasted three times longer than usual. The Chinese wrote of an incredible time when the sun did not set for several days while the land burned. And just as the buildings and works of men crumbled and toppled, so too did their institutions. People and whole nations were torn from their lands and set to aimless wandering.

The catastrophe also was responsible, according to Velikovsky, for the most memorable drama in the Old Testament: the Exodus of the Jews from Egypt. The awful devastation toppled the Egyptian Middle Kingdom, and Moses led the people of Israel, erstwhile slaves, out of the ruined land. As they fled across Egypt, before them, shining through the clouds of smoke and dust, hung the huge, dragonlike tail of the comet. At night the sky alternately glowed with the light and dazzled with the arcs of brightness as the comet's head and writhing tail fired huge electrical bolts back and forth.

The unceasing exchange of huge lightning bolts, electrical discharges that leaped explosively between the head and tail of the comet, seemed to the ancient peoples, already numbed by the series of incredible disasters that had befallen them, a battle between gods, monsters or demons. The Babylonians recorded the event as the titanic battle between the god Marduk and Tiamat, the dragon, fought with hurled lightning bolts. To the Egyptians it was the violent war of Isis and Seth; to the Hindus, the battle between Vishnu and the "crooked serpent"; to the Greeks, Zeus locked in combat with the coiled viper Typhon.

For the fleeing Israelites it marked the way to the Pi-Hahiroth—the Sea of Passage. Behind them raced the angry and vengeful Pharaoh Taoui-Thom and his mighty army. Ahead the seabed lay uncovered by the shifting movements of the earth's crust and the gravitational pull of the comet, its waters piled high on either side. The Israelites hesitated and then rushed to cross the sea bottom; behind them Egypt, ahead the Sinai Desert. With many of the fleeing people still in the seabed, two things happened almost simultaneously. The pharaoh's armies entered the sea, and the comet moved closer to the earth, as close as it was to come. An incredibly powerful electrical bolt passed between the earth and the protoplanet. The walls of water collapsed in a boiling, seething caldron, smashing and destroying the Israelites still in the seabed and the entire army of the pharaoh.

For weeks thereafter Venus punished the earth, flinging electrical bolts at the larger planet, searing and torturing the earth. For the Jews the devastation was a dangerous path to freedom marked by a pillar of smoke by day, a column of fire at night. And as the Israelites moved through the desert, following the alternately blazing and smoking marker that promised them salvation, the earth, too, continued moving through the comet's tail. For years the earth was shrouded with thick, choking haze, a gloomy pall that marked the end of a historical age.

Throughout the world handfuls of people survived where once thousands had lived, and even these pitiful few were threatened with starvation, for little in the way of food—animal or vegetable—had survived. And then yet another miracle was recorded in the Old Testament:

manna. The hydrocarbons in the comet's tail that had saturated the earth with burning naphtha and drenched it in petroleum were now being slowly changed within the earth's atmosphere, possibly by bacterial action, possibly by incessant electrical discharges, into an edible substance —manna. It fell as the morning dew, a sweet yellow hoar-frost filling the air with a delightful fragrance and the bellies of the hungry with nutrients. When heated, the heavenly bread dissolved, but when cooled, it hardened into grains, which could be stored for long periods of time and then milled between stones. It made survival of people and animals possible.

"But," Velikovsky points out, "if the manna fell from the clouds that enveloped the entire world, it must have fallen not only in the Desert of Wanderings, but everywhere; and not only the Israelites, but other peoples, too, must have tasted it and spoken of it in their traditions."

And so, of course, they did, and he cites the legends of Iceland, India, Greece and elsewhere around the world that speak of ambrosia, manna, madhu and sweet dew, dropping from the heavens to the earth.

But far more destruction than benefit resulted from the earth's passing through the comet's tail. The close approach of the comet produced gravitational dislocations that reversed the direction of the earth's axis. To the shocked and dazed people of the earth it appeared that the sun was now rising in the west and setting in the east. The fifth-century B.C. Greek historian Herodotus quotes Egyptian priests, who in turn cited their own ancient records of catastrophes: "Four times in this period (so they told me) the sun rose contrary to his wont; twice he rose where he now sets, and twice he set where he now rises."

Seasons were exchanged. "The winter is come as summer, the months are reversed, and the hours are disordered," states an Egyptian papyrus. In China the emperor sent scholars throughout the land to locate north, east, west and south and to draw up a new calendar.

For a generation the earth was shrouded in an envelope of clouds—the shadow of death of the Scriptures, the *Götterdämmerung* of the Nordic races. It endured for twenty-five years according to Mayan sources. The Israelites who had escaped the whirlpool in the Sea of Passage wandered through the Sinai Desert for forty years until, finally, they entered the land of Canaan behind a new leader, Joshua. Here they began a series of battles with the Canaanites that lasted for years. Then, as the Canaanites fled the army of Joshua through the valley of Beth-Horon, Venus again exerted its dreadful effect upon the affairs of men. Above Beth-Horon the sun appeared to stand still and the heavens to rain boulders upon the fleeing Canaanites. "The Lord cast down great stones from heaven upon them unto Azekah and they died." The terrestrial axis again tilted, and the earth heaved and buckled. Cities flamed up and collapsed in heaps of ash. On the other side of the world the Mexican records speak of a lengthened night illuminated only by the burning of the land about them. This, they noted, occurred some fifty years after a previous catastrophe.

And once again the earth was racked and sundered by earthquakes, global hurricanes, continental shifts and universal destruction. But the planet and some of its inhabitants endured. To the survivors who had suffered through two catastrophes in two and a half generations, a pattern became evident, and the peoples of the world, each in a manner dictated by their own culture, bowed down before the dreaded Venus, the goddess of fire and destruction. With human sacrifices and bloody rituals, with prayers and incantations they pleaded with the awful queen of the heavens to leave the earth in peace. Cuneiform tablets found in the ruins of the library palace in Nineveh, the Assyrian capital, record the erratic behavior of Venus. "How long wilt thou tarry, O lady of heaven and earth?" inquired the fearful Babylonians. "We sacrifice unto Tistrya, the bright and glorious star," mumbled the Persians, "whose rising is watched by the chiefs of deep understanding."

For a time the prayers were answered as Venus blazed along its circuit between the orbits of the earth and Mars. And there appeared to be a relative form of order restored to the planets as the first series of catastrophes came to an end, according to Velikovsky.

While the ideas contained in *Worlds in Collision* fit very neatly the extensive collection of myth, legend and literature Velikovsky amassed to prove his point, he also contended that there were three crucial tests concerning Venus that would prove or disprove the theory.

The first of these crucial tests centered on the surface

temperature of Venus. At a time when the codified body of astronomy uniformly insisted that Venus was a relatively cold planet, Velikovsky wrote: "Venus experienced in quick succession its birth and expulsion under violent conditions; an existence as a comet on an ellipse which approached the sun closely; two encounters with the earth accompanied by discharges of potentials between these two bodies and with a thermal effect caused by conversion of momentum into heat; a number of contacts with Mars, and probably also with Jupiter. Since all this happened between the third and first millennia before the present era, the core of the planet Venus must still be hot."

In December 1962, twelve years after *Worlds in Collision* was published, the U.S.-launched space probe Mariner 2 flew past Venus and recorded a surface temperature of 800 degrees F, 200 degrees above the melting point of lead. More recent and exact tests have increased that measurement to more than 900 degrees. Despite thousands of years of cooling off, Venus is indeed hot.

But Mariner was to reveal still more that would confirm the accuracy of another of the ideas that naturally followed from Velikovsky's work—anomalous rotation. All planets, according to the astronomical dogma of the day, were thought to rotate in a counterclockwise direction. That is, if you were to look down upon each planet from above its north pole, it would appear to be spinning in a direction opposite to the hands of a clock.

Radar measurements of Venus made by the tracking station in Goldstone, California, in 1964 as Mariner 2 was passing the planet, however, established finally and conclusively that Venus possessed a retrograde rotation.

Yet another of Velikovsky's predictions concerning Venus, which he considers crucial in testing his theory, concerned its atmosphere.

"I have claimed a massive atmosphere around Venus," says Velikovsky, "while my 1951 reviewer and opponent, the Royal Astronomer Sir H. Spencer Jones, maintained that Venus has less atmosphere than the earth.

"In 1967 the Russian probe, unprepared for the enormous pressure it encountered, was crushed while descending toward the surface of Venus. The Venusian atmosphere it turned out is 95 times as heavy as that of earth."

But it is in the chemical composition of that Venusian atmosphere that the ultimate test of the Velikovsky theory may be made. "I also claimed," says Dr. Velikovsky, "that in historical times the trailing part of the protoplanet Venus became partly absorbed into the atmosphere and cloud covering of Venus and that quite probably till today there are hydrocarbons present or, instead, quite possibly organic molecules." The hydrocarbons Dr. Velikovsky talks about are petroleum products which are composed of only two elements—carbon and hydrogen. The organic molecules he talks of, which are the source of the manna that fell to earth, are composed of carbon, hydrogen and oxygen. Thus, the atmosphere of Venus would have to have these elements present, either combined together as hydrocarbons and carbohydrates or at least as individual molecules capable of joining together to make those products.

It was hoped that Mariner 10 might provide answers to this prediction as it scanned the atmosphere of Venus. (This was not an avowed goal of Mariner 10, and in fact, not one of NASA's principal scientists is examining the data with any idea of applying them to Velikovsky's theory, but these data nonetheless can be used to examine the requirements of the theory.)

"We've seen an extended atmosphere of atomic hydrogen which surrounds Venus," said the principal investigator, Dr. Michael McElroy of Harvard. "We have seen helium for the first time. We're seeing the atom carbon; we're seeing a fairly extended atmosphere of carbon which scatters sunlight and which is quite bright. That's new. We've seen atomic oxygen. And we've seen lots of carbon monoxide in the upper atmosphere of Venus; somewhat surprising, despite the fact that the planet's atmosphere is made of carbon dioxide."

Carbon, hydrogen, oxygen—all the elements needed to make hydrocarbons and carbohydrates have now been definitely identified in the upper atmosphere of Venus, but not the actual compounds themselves. Does that mean that Dr. Velikovsky is wrong?

Not necessarily. What Mariner has revealed is the chemical composition of only the highest clouds of Venus. "We are looking," Dr. McElroy points out, "at little slices of the planet, which are very, very thin, on the order of 20 kilometers in vertical extent."

Deeper down, closer to the surface, the individual carbon, hydrogen and oxygen molecules might be combined into hydrocarbons and carbohydrates, as Velikovsky suggests. "If Venus had a rather rapid evolution, if it was a hot incandescent body as Velikovsky says, and then everything happened very, very fast, over the space of a few centuries then what we see today could be quite different than when it was a young protoplanet," says Dr. Albert Burgsthaler, professor of chemistry at the University of Kansas, who began debating the chemistry of the Venusian atmosphere with Velikovsky in 1970. "The basic materials are there, there is an awful lot of carbon there from which to make hydrocarbons and if he's right, we've had a great deal more chemistry take place since Venus settled down into orbit as a planet."

Basic chemistry is at the heart of the most dramatic miracle of that period, the fall of manna from the heavens. In 1951, in a response to his critics, Velikovsky speculated that "through prolonged bacterial action, hydrocarbons could have been converted into edible products."

The idea was dismissed by scientists as absurd. But today there are at least half a dozen experimental plants in the United States and Europe that are utilizing bacteria to convert hydrocarbons in the form of crude oil into a food substance that contains more protein than beefsteak.

Other processes that might produce manna without the use of bacteria also been demonstrated, and the creation of amino acids, the building blocks of protein, from a primal mixture of ammonia, methane, water vapor and hydrogen, was accomplished in 1952 by one of Velikovsky's severest critics, Dr. Harold Urey, then a professor of chemistry at the University of Chicago. Water vapor and hydrogen have been found on Venus by the Mariner space probe, and the Soviet Venera 8 probe of Venus detected the presence of ammonia. Methane, the fourth substance in the Urey mixture, is the simplest hydrocarbon, formed by joining four hydrogen atoms to a single carbon atom, both elements found in abundance on Venus.

Mariner 10 found another attribute of Venus that adds still more credence to the Velikovsky concept of the planet's younger career as a comet. "Unusual intermittent features observed downstream of the planet," reported the scientific team studying the results of Mariner's plasma science experiment, "indicate the presence of a comet-like tail hundreds of scale lengths in length."

Nor is Venus the only planet that should be explored in seeking evidence of its violent birth and travels through the solar system. After its first two destructive passages past the earth Venus continued its blazing course across the orbits of both the earth and Mars. People noted its passage with great care and fear, and toward the middle of the eighth century B.C. astrologers could see dramatic changes in its path. The Babylonians recorded on their tablets the disappearance of Venus in the west for more than nine months and then, with equal suddenness, its reappearance in the east. Then it dropped below the horizon, vanishing from sight for two months, when it suddenly rose in the west. The following year the tablets recorded the disappearance of Venus into the west for eleven days before it once more blazed forth in the eastern sky.

The erratic behavior of Venus frightened the planet watchers of earth, but the immediate threat was to Mars. "Mars, being only about one-eighth the mass of Venus, was no match for her." Mars suffered a cosmic dislocation when Venus brushed past. The smaller planet was pulled out of its orbit and flung onto a new path about the sun, one that threatened the earth.

As Mars drew near, the earth staggered in its orbit. Once again, cities collapsed, earthquakes sundered and split the surface, and men died amid flaming geophysical changes that occurred within instants. The prophets Isaiah, Hosea, Joel and Amos recorded these catastrophes, and they are also described in Homer's *Iliad*.

Just as the previous cosmic disasters altered the course of history, ending a historic age and marking the beginning of a new one, so too did the appearance of Mars produce great changes in the affairs of men. The cataclysms were accompanied by the founding of Rome in 753 B.C. "Both the poles shook and Atlas shifted the burden of the sky," wrote Ovid. "The sun vanished and rising clouds obscured the heaven. . . ." And Rome had a powerful god to appease—Mars, the god of war.

The effects of the close passage of Mars could not equal those of Venus, for it was so much less massive than the earth. But the gravitational pulls were great enough once more to shift the earth in its orbit about the sun. The old calendar of 12 months of 30 days each, adding up to 360-day years, was no longer accurate. Pushed into a wider orbit, the earth took a longer time to traverse its path about the sun, and emperors and kings ordered their astrologers to devise a new calendar, one that would acknowledge the new physical realities of the earth's path about the sun.

Mars and Venus now became the chief deities in most cultures, for they determined the fate of nations. Wars were fought in the name of Mars or Venus, cities and temples were dedicated in their honor, and all human affairs were dictated by the effects of these two planets on the earth. Seers, prophets and astrologers continued to prophesy calamity and doomsday. Eighth-century Hebrew prophets reminded the Israelites of their exodus from Egypt and said that once again the earth would be shaken by earthquakes, the moon turned to blood, the sun shadowed and the land tormented by fire and smoke.

In 721 B.C., on the day King Ahaz of Jerusalem was buried, Mars roared past the earth. The results were once again catastrophic as the earth's axis tilted and the poles shifted. Its orbit swung wider, lengthening the year. At the moment of closest passage, when the gravitational attractions of both bodies were greatest, the sun as viewed by the Israelites seemed to hurry to a premature setting, dropping below the horizon several hours before it normally set. Thereafter an observer in ancient Israel would see the sun move across the sky 10 degrees farther to the south.

The Greeks noted the same effects, the philosopher Seneca recording an early sunset accompanied by a great upheaval on the Argive plain. Then, at night, the Great Bear constellation dipped below the horizon; familiar constellations changed their positions in the days that followed. "The Zodiac," wrote Seneca, "which, making passage through the sacred stars, crosses the zones obliquely, guide and sign bearer for the slow moving years, falling itself, shall see the fallen constellations."

Although Mars did far less damage than Venus had some seven centuries earlier, it became a dominant and fierce god in the pantheon of man's heavenly forces. Velikovsky believes that its last cataclysmic approach took place in the spring of 687 B.C. In that year the Assyrian King Sennacherib marched against Hezekiah, a king of Judah, planning to capture Jerusalem. On the evening of March 23, the first night of the Hebrew Passover, Mars made its last terrible approach past the earth, unleashing a great thunderbolt, "a blast from heaven," that charred the bodies of the besieging soldiers but left their clothes untouched. Sennacherib fled in horror, leaving 185,000 of his army dead before the walls of Jerusalem.

The same night the Chinese recorded a disturbance of the planets that caused them to go "out of their courses. In the night," said the *Bamboo Books*—and they give the date—"stars fell like rain. The earth shook." French scholars calculated that the event took place on March 23, 687 B.C.

To the Romans, March 23 became the festival of *Tubilustrium,* a major celebration in honor of the god of war, Mars.

In some longitudes, as Mars made its last terrible pass at earth, the rising sun dipped back below the horizon. In others, the setting sun popped back up above the horizon. This retreat of the sun was caused by tilt in the earth's axis, a tilt that nearly corrected the one that had occurred a generation earlier.

"So the sun returned ten degrees, by which degrees it was gone down on the Sun Dial of Ahaz," recorded Isaiah.

To the terror-stricken watchers on earth, it seemed that Mars and Venus were battling each other in a cosmic arena and that wildly thrown thunderbolts from that battle threatened the bystander, earth. Homer recorded the battle in the *Iliad.* As the Greeks besieged Troy, Athena (Minerva) "would utter her loud cry. And over against her spouted Ares [Mars], dread as a dark whirlwind. . . . All the roots of many founted Ida were shaken, and all her peaks." The river "rushed with surging flood," and the "fair streams seethed and boiled."

Neither Mars nor Venus was to triumph. Mars was cast back out beyond the range of danger to the earth. And Venus, which had assumed a dominant role in the heavens, soaring up to the zenith, dropped back to become a morning and evening star that never rises to zenith.

As the other major planet caught in the wild career of Venus through the heavens, Mars must also retain some evidence of the events. Velikovsky thought so and predicted it would have, at least partly, a moonlike surface, cratered and pitted.

And so it has turned out. To the surprise of astronomers, pictures sent back to earth by probes show a rough surface, pocked and cratered with remnants of huge volcanoes, large faults and rifts, valleys and channels that could be only the products of some massive geological stress.

Velikovsky had also claimed since 1945 that the rare gases argon and neon would be present in the Martian atmosphere in rich amounts. Some astronomers had also been predicting the presence of argon. In 1974 Russian probes to Mars detected a substantial amount of argon and neon in the rarefied Martian atmosphere.

Most of the astronomical confirmations of Velikovsky's theories have come from long range, relayed through the mechanism of robot explorers, but in the case of one planet, man himself has been the explorer. From July 1969 through December 1972 a remarkable series of six spaceflights placed twelve men, machines and instruments upon the moon and brought back more than 600 pounds of rocks and dirt from various sites on the lunar surface. Left behind were a variety of experimental instruments that continue to send back still more information about the moon's surface and interior.

The amount of information brought back from the moon is staggering. It demolished several long-cherished ideas and established beyond doubt that the moon and the earth were created separately and have evolved differently. But the data also raised almost more questions than they answered, and much about the moon remains as mysterious as before Apollo 11 landed the first men on the moon.

Astronomers and lunar geologists, for example, have no way to explain the wholly unexpected presence of magnetism in the lunar rocks. Yet before Neil Armstrong stepped off the "front porch" of the Apollo 11 lunar module onto the surface of the moon, *The New York Times* carried an article by Velikovsky which predicted the lunar rocks would be magnetized.

"The moon has a very weak magnetic field," he wrote, "yet its rocks and lavas could conceivably be rich in remanent magnetism resulting from strong currents when in the embrace of exogenous magnetic fields."

The moon could have encountered such exogenous magnetic fields if it has been approached by other celestial bodies, as Velikovsky has claimed.

This, of course, is an explanation astronomers refuse to credit. And for them, the question of how the lunar rocks acquired their appreciable quantities of remanent, or residual, magnetism remains unanswered. "The origin of these fields in which remanence was initially acquired is the central unsolved question in lunar magnetism studies," admitted the Lunar Science Institute in its report on *Post-Apollo Lunar Science* in July 1972.

Velikovsky made a number of other predictions concerning the moon, all of which have proved to be true. He predicted, for example, that the moon had been subjected to heating and bubbling activity in recent times. Investigators examining the lunar finds—the gravel brought back from the moon—found that a large part was composed of tiny glassy spheres. Naturally occurring glass is extremely rare on earth and is the product of high pressures and great heat. Its presence can be used to justify Velikovsky's claim that the moon had been heated when Mars passed close by.

NASA scientists offer a different explanation of the presence of so much glass on the moon. They claim it is the result of meteorites' striking the moon, which produces the requisite heat and pressure.

The lunar geologists and Velikovsky, however, are in agreement on the bubbling activity that took place on the moon. An extremely common feature was the vesicles, or holes, that riddle most of the lunar rocks brought back to earth. The holes were created by gases that bubbled through the rocks when they were still liquid.

Where Velikovsky and the lunar scientists part company is on the question of when these events took place. For although the physical evidence of the melting and bubbling is present in the rocks, the radioactive clocks used to date them offer wildly differing ages. Among the means of dating ancient rocks is to examine them for radioactive elements that change slowly at known rates into other elements. Potassium, for example, decays into a radioactive isotope of argon.

Before the first rocks were returned from the moon, Velikovsky predicted they would contain large amounts of both argon and neon. He also told Professor Burgsthaler that the presence of these gases would be misinterpreted. "It is quite probable," he wrote, "that some confusion will be created in the minds of many when on the basis of potassium-argon or some similar reaction, the age of the rocks will be established in billions of years."

Six months later the journal *Solar Science* reported: "One of the most puzzling aspects was the high concentration in the rocks of Argon 40. That isotope is produced routinely by the decay of potassium, but the excess abundance of the isotope is so great that it would imply an embarrassingly high potassium-argon age for the moon—about 7 billion years, a figure no one takes seriously."

Other reports offered ages as high as 20 billion years old, older by several billion years than the universe itself. Any age over 4.6 billion years—the generally accepted age of the solar system—is suspect, and any dating system that gives figures of almost twice the age of the moon itself is even more suspect. But, as Dr. Velikovsky points out, "the question is not *when* the rocks have been *formed* or for the *first* time crystallized, but when they were heated and partly molten for the last time. The *age* of the *rocks* is not in dispute, only the time of the 'carving' of the lunar surface." And indeed most of the lunar rocks were dated by other means at from 3 to 4 billion years old.

The last "carving" of the rocks took place, according to Velikovsky, 3000 years ago, when Mars passed close by the moon as part of the cosmic disruptions begun by Venus after it had been ejected from Jupiter.

Despite all the denials, the charges of crackpot and scientific heretic, the power and logic of the Velikovsky argument remain undiminished. As we search ever deeper into space and dig with more skill into the record of our own past on the earth, Velikovsky finds more evidence supporting his thesis. Meanwhile, interest in the ideas expressed in *Worlds in Collision* continues to grow. In a world increasingly sensitive to ideas and now thoroughly conditioned by the media, the continuing stream of reports from distant planets and dusty desert digs becomes a welcome change from the ever-present news of war and strife around the world. And a theory that seems to make sense out of our cosmic probings and our own convoluted history becomes all the more appealing.

The continuing interest in the work of Immanuel Velikovsky can perhaps be ascribed in some measure to that rationale. Since its first publication in 1950 *Worlds in Collision* has gone through seventy-two editions in the English language alone. Velikovsky's theories have become the subject of courses in at least forty universities, and a number of scholarly magazines, such as the *New Scientist* and *Yale Scientific,* have devoted articles and even entire issues to his ideas. One tiny political science quarterly called

*Pensée,* published for the students in the Oregon University system, blossomed into an international journal of no small fame when it first devoted an entire issue to the single theme "Velikovsky Reconsidered." For ten "Velikovsky" issues, *Pensée* was a forum for debate and discussion of the Velikovskian ideas, with both supporters and detractors publishing a range of articles dealing with everything from celestial mechanics to the significance of ancient pollen grains found in the waters off Pylos.

With this new interest, a wave of seminars and symposia has been held, including what was to have been Velikovsky's ultimate recognition by the scientific establishment, a symposium held by the American Association for the Advancement of Science, the world's largest scientific organization. It was hardly a triumph as scientists, quoting formulas, citing physical laws and involved mathematics to fit their own arguments, demonstrated why Velikovsky was wrong. It was not a hearing, but another judgment rendered by organized science, but this time the ideas were not dismissed out of hand. Rather, a careful attempt was made to put an end at last to the theories that keep nagging at the established rules that science claims govern the solar system.

But no sooner had the symposium ended and all participants concluded that the Velikovsky theory had finally been put to rest than the results of Pioneer 10, the Jupiter probe, and Mariner 10, the flyby past Venus and Mercury, began to come in. And as had happened in the past, with previous explorations in space, information appeared to support Velikovsky.

After twenty-five years we seem to have come almost full circle. The ideas expressed in *Worlds in Collision* hold as much appeal and power as the day they were published, and the scientific establishment continues to deny their author even the elemental courtesy of acknowledging that he has at least made some astonishingly accurate predictions about the physical states of Jupiter, Mars, Venus and the moon.

The record of Velikovsky's predictions does point up the unyielding, unremitting refusal of the scientific establishment to accord him even the slightest hint of legitimacy. But just as astronomers and geologists hold that human history is but an eye blink when measured against the grand time scale of nature, so, too, may their objections to an eighty-year-old scholar's theory be seen as but a minor blemish on the long history of human ideas.

"It really does not matter so much what Velikovsky's role is in the scientific revolution that goes now across all fields from astronomy, with emphasis on charges, plasmas and fields, to zoology with its study of violence in man," wrote Velikovsky. "But this symposium in the frame of the AAAS is, I hope, a retarded recognition that by name calling instead of testing, by jest instead of reading and meditating, nothing is achieved. None of my critics can erase the magnetosphere, nobody can stop the noises of Jupiter, nobody can cool off Venus and nobody can change a single sentence in my books."

# 4

# MYTH—EYEWITNESS TO CATASTROPHE

---

It is not the purpose of this book to deny the established truths of science. There are laws that can be discerned in the workings of the universe. The broad body of science does not require *deus ex machina* to explain the unexplainable. Where natural phenomena occur without our understanding, it simply means we do not yet know enough, not that there is necessarily some convenient cosmic visitor to account for the mystery. Rather, what I am suggesting is that there are additional rules, clearly visible in the phenomenological evidence, that we have largely ignored in our headlong search for "pure truth."

Thus, it has been with catastrophism. Most of its adherents attempted to fit it within the time frame of the Bible, and when certain biblical dates were ascertained and placed alongside geological time, the entire edifice of catastrophism collapsed. But the role of catastrophe in nature as an ultimately creative, rather than destructive, force remains the same. Still, catastrophism is considered a relic of a bemused past when allegory was taken literally, when science was the province of amateurs and truth less sought after than theologically comfortable beliefs.

So, too, has it been with myth. *Webster's New Twentieth Century Dictionary,* the one that takes two people to lift, defines "myth" as: "a traditional story of unknown authorship, ostensibly with a historical basis, but serving to explain some phenomenon of nature, the origin of man, or the customs, institutions, religious rites, etc. of a people; myths usually involve the exploits of gods and heroes."

There are other definitions, with subtle nuances, but all tend to put down the historic validity, just as Mr. Webster's lexicographers do with the phrase "ostensibly with a historical basis." The idea of myth as history in this day of lunar landings and space probes to the very borders of the solar system is somehow degrading. Modern man has a compulsion to solve mysteries. Myth simply records them. The authors of myth, presumably people less scientifically sophisticated than we are, did not require that their mysteries be solved. They were simply to be believed, nothing more, for they were reports of events that had happened and were to be recorded for all the generations that were to come.

Today, however, we deny whatever truth myth may hold. Hundreds of books by learned scholars and scientists have been written about myth; many purport to explain myth scientifically, yet virtually all aim instead at denying the historic validity of most myths. That ancient man might have been an accurate observer and recorder of natural phenomena seems to be an idea modern man cannot easily accept. We insist on seeing in primitive man a childlike, inept creature who, thanks to the happy accident of flint dagger and fire-hardened spear tip, managed to claw his way toward civilization. His mythology is seen as fable, as fairy tales to explain the terrors of the night, as shibboleths to kick a basically amoral group into observing a few of the more basic rules without which no stable society can exist.

And so myth became the cornerstone upon which religious beliefs were erected. Myth we now view as allegory, as storybook fantasy created to answer the questions and calm the fears of a primitive, childlike people. From such mythical forms, ultimately the ethical and moral edifice of religion was erected. Or at least that is what we have always been led to believe.

"Religious thought does not come in contact with reality," wrote sociologist Émile Durkheim in his famous study of primitive religion *The Elementary Forms of Religious Life*, "except to cover it at once with a thick veil which conceals its real forms: this veil is the tissue of fabulous beliefs which mythology brought forth. Thus, the believer, like the delirious man, lives in a world peopled with beings and things which have only a verbal existence."

Durkheim is, of course, not alone in this view. Concurrent with the rise of uniformitarianism has been the highly successful effort by scholars to deny the historicity of the events depicted. They are instead dismissed as fables, and the psychological and existential messages beneath the myths are then sought.

But even the nineteenth-century uniformitarians had their precedent—and in the church itself—for denying the historical validity of biblical myth. In the third century the priest Origen declared that some passages in the Bible "are not literally true, but absurd and impossible." Two centuries later St. Augustine counseled against interpreting the Bible literally when he wrote, "We must be on guard against giving interpretations of Scripture that are far fetched or opposed to science."

This same viewpoint today, whether admitted or not, is the chief objection to the Velikovsky view of the Exodus. Despite the scientific evidence supporting his viewpoint, he committed the unpardonable sin of basing his thesis on the myths of the Hebrews and dozens of other peoples about the world who all told of incredible disasters befalling the earth. That their viewpoints agreed with their geographic locations mattered not one whit, that Venus is hot and space is filled with electromagnetic fields, in the view of most scientists and theologians are still not evidence that this particular myth is, in fact, history.

For a startling change has come about in religious and theological belief. Where once catastrophism was used to support the events of the Bible, now those events are denied as fact unless they support science. Theologians in the exegesis of biblical accounts are at great pains to point out that stories such as Exodus are merely symbolic in nature, that they did not actually happen but are merely used to glorify and romanticize a people and its history.

"It is obvious to any unbiased reader that this story, with its markedly religious coloration and its emphasis on supernatural 'signs and wonders,' is more of a romantic saga or popular legend than an accurate record," wrote Jewish scholar Theodor Gaster, of the Exodus.

The viewpoint of Christian scholars that the Exodus

never actually happened is no less certain. In *Israel: Its Life and Culture* the Swedish Protestant theologian Johannes Pederson declares: "In forming an opinion of the story about the crossing of the Red Sea, it must be kept in mind, as we have remarked above, that this story, as well as the whole emigration legend, though inserted as part of an historical account, is quite obviously of a cultic character, for the whole narrative aims at glorifying the god of the people at the paschal feast through an exposition of the historical event that created the people. The object cannot have been to give a correct exposition of ordinary events but, on the contrary, to describe history on a higher plane, mythical exploits which make of the people a great people, nature subordinating itself to this purpose."

What a colossal impertinence on the part of Pederson. How can any scholar with any pretensions to scholarship so twist an account of an event that it becomes merely a propagandistic attempt to glorify the group at the expense of nature? But Pederson's interpretation is typical of the denial of myth as history or as evidence of catastrophe. Even where the Exodus is accepted as a historic event, its magnitude is vastly diminished. The parting of the Red Sea is not the product of some vast global calamity, but of a far lesser phenomenon. The Red Sea is transmogrified into a rather shallow lake or lagoon, and the Israelites splash through when a high wind springs up and piles the shallow waters upon the banks.

But suppose that the Red Sea was indeed a great body of water and that it was indeed parted by a violent windstorm that was the product of some vast global catastrophe? Suppose, in other words, that Velikovsky is right?

"Suppose Immanuel Velikovsky is correct?" asks Vine Deloria in his book *God Is Red*. "Suppose that instead of the Exodus accounts being a poetic elaboration of religious doctrine of a later time, they are fairly well remembered accounts of the phenomena encountered by the Hebrews as they left Egypt. How then do we approach religious writings? Are they to be understood as actual events and do we take all religious stories as having been real events at some time and some place in man's experience? It would seem that we have a major task of discovering to what extent we can accept the historical veracity of any story of ancient times. The fact that Immanuel Velikovsky's projections about the nature of the physical world continue to produce startling verifications would tend to make us back up and take another look at religious doctrines, symbolism and belief."

Nowhere is that belief more strained, is credibility more questioned, than in the first ten words of the Bible. "In the beginning God created the heavens and the earth." This is the quintessential creation myth, echoed and retold in hundreds of other cultures, but with only slight variations. But it cannot be historic. It cannot be an eyewitness account of a natural phenomenon, for man could not have possibly witnessed it. If we accept even the barest outlines of evolution, that life has evolved on earth from a handful of chemicals through a series of stages that had led from the single cell to the blue whale, with intermediate and penultimate forms along the way, including man, then man, the maker of myths, could not actually have witnessed the creation. What he could have done was to draw on his primitive understanding of nature and develop a story to explain the event. This, of course, is a perfectly logical conclusion and, in fact, fits the rules of this particular ball game where catastrophe is the home run ball. It does, of course, credit ancient man with far more perception and understanding of natural processes than we have ever given him credit for, but after all, we never denied his ability as a storyteller.

So be it. Now listen to another story.

A 1973 meeting of the American Chemical Society took up a problem encountered by the Apollo astronauts. They had complained about sticky dust clinging to their boots while walking on the moon. The explanation was simple. Bombarded by cosmic rays, dust particles become electrically charged. On the moon they stick to astronaut boots. In space they stick to each other. In this manner, pointed out Dr. Gustav Arrhenius of the University of California, they form larger particles in which organic chemicals might be created. In a cloud condensing into a star such as our sun, some dust clumps could grow into the nuclei of planets. "Thus," said Dr. Arrhenius, "they could bring with them the starter chemicals from which life could arise on such planets."

How, save in degree, does this theoretical model of the creation differ from the biblical account?

"And the earth was without form, and void: and darkness was upon the face of the deep. And the Spirit of God moved upon the face of the waters. And God said, Let there be light: and there was light."

Is the concept of starter chemicals for life any less fanciful than the biblical account which declares: "So God created man in his own image, in the image of God created he him; male and female created he them"?

Did the science of the nomadic desert tribe we know as Hebrews enable them to deal with dust clouds that condensed in darkness until they reached the critical temperature and pressure that would ignite their nuclear furnaces and enable them to blaze forth as stars? "Let there be light." Within the framework of a desert experience we can see the physics of creation.

This, of course, is physics, not history. Are we to accept, then, the observational and deductional abilities of ancient man and discount him as a witness to catastrophe? Is he guilty of snowing us with his purple prose and poetic descriptions of disaster and calamity that never quite happened? That is doubtful.

Not all myth is required to be actual eyewitness depositions, and there are, of course, elements of the dictionary definition in some of the world's folklore and legend. And certainly embellishment of events by preliterate storytellers would be the rule rather than the exception, but this does not rule out man as a keen observer of natural events, especially when they bring the world crashing down about his ears. Even the creation myth can be posited as an eyewitness account, if it were a *re-creation* myth.

Recorded history reaches back only 6000 years into our past. Yet anthropologists tell us that man may be as much as 4 million years old. That leaves 3,994,000 years unaccounted for. Did nothing untoward occur during that time? Did the generations of human beings approach their present preeminence with nothing more to worry about than an occasional saber-toothed tiger, cave bear, plague or war?

Obviously not, and the remarkable accounting of catastrophic terrors is, in fact, retold as myth and legend. During the 3- or 4-million-year span when man was clubbing his way up from the bottom of the Olduvai Gorge to the top of the World Trade Center, he obviously was subjected to a host of disasters. If Velikovskian catastrophe can brush the earth four times within recorded history, how many more times can global disaster strike, during the course of 3,994,000 years? Could the stable, slowly evolving earth have been less than stable, less than uniform in its evolution? Could recurrent catastrophes, crashing continents, cometary brushes, earthquakes, tidal waves, volcanic eruptions not have wracked the earth at various times during man's tenure on its surface? Why do we have records that go back only 6000 years when our anthropological record goes back 4 million years? Is it possible that the record of man's intellectual development is one of startling and sudden leaps? How was it that such human achievements as the calendar, agriculture, animal husbandry, art, writing and science developed so suddenly with no evidence of the steps leading up to them?

Could the record have been there and been totally and completely erased?

Once upon a time the library at Alexandria was considered the ultimate repository of knowledge. All the ancient world's wisdom was deposited there on papyrus and clay tablet. All of it vanished in a great fire, 2000 years ago, never to be seen again. The same process of catastrophe might have erased records that had been written not 2000 years ago, but 20,000 or more years ago. If the devastation were complete enough, these records would not have survived.

Among the earliest forms of written record or notation is the calendar. The official Egyptian calendar dates from 4000 B.C. But in 1972 Alexander Marshack, a journalist turned archaeologist, published the results of the ten years of research he had done on a scratched Mesolithic bone. The bone was 8500 years old, and the inescapable conclusion was that the scratches and marks on it were not an abstract decoration, but a sophisticated and accurate lunar calendar.

True, calendric notations are not writing in the classic sense, but they do bespeak an enormous intellectual sophistication based on language communication. And where there is language there is soon some form of literacy. But there is more, much more to the possibility that man was able not only to see and understand what was happening in his world, but also to record it far earlier than we ever believed before.

At a recent meeting on the origins and evolution of language at the New York Academy of Sciences, Dr. Marshack, now an archaeologist at Harvard's Peabody

Museum, displayed photographs of yet another scratched bone. They showed a piece of ox rib dug up in France in 1969 and dated at 300,000 years. On the bone someone had scratched pairs of parallel lines that formed a running zigzag over the surface.

According to Dr. Marshack, the zigzag lines seemed to be symbols. They were identical to the zigzag motif that covered many artifacts left by prehistoric man. "I assume," he told the assembled linguists, anthropologists and archaeologists, "that the images required some form of spoken language to explain and maintain the tradition."

So 300,000 years ago a smaller-brained ancestor of modern man, known as *Homo erectus,* was probably using language and making notes about his world.

The possibilities of language and understanding having developed even earlier are pushed to astonishing limits inside the cranial bones of a fossil skull known as 1470. Discovered in Kenya by Richard Leakey in 1972, it is the remains of a creature that lived between 2 and 3 million years ago. Possessed of a smaller brain than *Homo erectus,* he nonetheless may have been able to use language. Not merely to communicate in barks and grunts, mind you, but to utilize language in the form of words and phrases that made complex understanding between two creatures possible.

At that same meeting in New York Dr. Ralph Holloway, of Columbia University and a specialist on the evolution of the brain, offered evidence of a possible language ability by the owner of skull 1470 when he was alive. Casts were made of the interior of the empty skull 1470. The casts then were removed and became in effect approximate reproductions of the brain that once filled skull 1470. On these casts Dr. Holloway found a barely discernible bulge that corresponded to a just visible bulge on the brain of modern man. That bulge is part of the region of the brain that is associated with language ability.

Dr. Holloway was quick to point out that his findings were tentative. "There is no good paleoneurological evidence that proves or disproves any language theory," he said. But other findings by other researchers tend to support the idea that the brain once housed in skull 1470 almost 3 million years ago might have been capable of speech.

What might they have said as they watched the earth

shaken to its very foundations? What tales and stories might the survivors of a great catastrophe have told? If after being rocked by a comet or having the sun wiped from the sky for a period of days or even weeks by a series of volcanic eruptions that filled the air with blinding ash or storms that went on for days and days, blackening the skies, flooding the earth and killing most of the animals and people who inhabited the earth, how would the survivors account for their salvation?

"What we have previously been pleased to call creation stories might not be such at all," says Vine Deloria. "They might be simply collective memories of a great and catastrophic event through which people came to understand themselves and the universe they inhabited. Creation stories may simply be the survivors' memories of reasonably large and destructive events."

The Hopi Indians have a mythology that depicts the destruction of the world on three separate occasions. Once it was destroyed by a rain of fire which a handful of Hopi survived by living underground with the ants. When they emerged, they might well have seen a darkened world suddenly flooded with light.

The Navajo also record a survival epic in which the first people emerged from the underworld. "Upon ascending into this world, the Navajo found only darkness and they said we must have light."

"The Navajos," explains Deloria, "then separate light into constituent colors of white, blue, yellow and black representing the colors of the sky during the twenty-four-hour period of rotation." Deloria offers similar versions of the same myth as shared by most of the Indian tribes of North America. "There would appear to be no reason for a number of tribes sharing this story, unless there was some event behind it even though dimly recalled in tribal memory."

The same might be said of all the other cultures of the world that share the creation myth. For if they were survivors of some global disaster, they would, in fact, be in a position to witness a rebirth, a re-creation of the world after the *Götterdämmerung* of catastrophe. How many times that happened, how far back in our racial memory such events are imprinted we may never know. All we can be assured of is that our memory is infinitely longer than we ever before acknowledged, that our ancestors, even those whose brain cases were smaller than our own, were far more intelligent and accomplished than we were ever prepared to admit, and from them comes more than a little truth about the catastrophic history of the earth.

Nor need we base our conclusion, our faith, if you will, in myth as history on such distant and admittedly speculative grounds. Consider that most explicitly detailed and endlessly recurring catastrophe myth, the great flood.

So strong a hold has it on man's imagination, thousands of years after it took place, if, in fact, it did, we still use words like "antediluvian" to speak of history before that point. Early geologists looking at the sand and gravel deposits far from the seacoast considered them the result of the biblical deluge and called them diluvium and named the time in which they were deposited the Diluvial. But later earth scientists determined that these deposits had been left by the glaciers of the Pleistocene time, and the idea of the flood as an actual universal event was dealt yet another blow.

It was not always so, for the flood tradition was so strongly rooted in most peoples' culture—not merely in the Judeo-Christian tradition—that it has the character of a "racial memory."

The Sumerians' tale of the deluge, as told in the Epic of Gilgamesh, is almost as well known as the biblical account. In this instance Noah is replaced by a man called Utnapishtim, and the stories of the ark building, the loading of family and cattle and beasts of the field and the eventual landing of the ark upon a mountaintop as the floodwaters receded are remarkably similar. That similarity, in fact, has led some scholars to insist that the biblical account is but a retelling of the Sumerian story. But there are other accounts of the deluge that are part of the mythic heritage of peoples so far removed from the Middle East as to be on the other side of the world.

All the Indians of South America have tales of a great flood which were told to the priests accompanying the conquering Spaniards. Cristóbal de Molina in his *Fables and Rites of the Yncas* sets down a myth of the people of Ancasmarca, a mountain in the Andes. Here a shepherd and his six sons and daughters assembled all the food and livestock they could at the top of Ancasmarca. They had been warned by their llamas of a coming flood. And indeed it came, but as the waters rose higher and higher to cover the earth, their mountain peak also rose so that it remained just above the waters. When the waters receded, the peak also grew lower. The shepherd and his six children then repopulated the land of Ancasmarca.

De Molina, after dutifully recording this myth, so astonishingly similar to the biblical account, went on to say, "These and similar follies they used to tell and still tell. To avoid prolixity, I omit them. Besides the chief cause of this, which was their ignorance of God and their abandonment to idolatry and vices, it was also due to the fact that they had no writing. If they had known how to read and write they would not have been addicted to such blind and stupid nonsense."

To the deeply pious and fundamentalist Spanish priests the deluge myths of the Indians could scarcely be conceived of as being a parallel tale to the one told in the Bible. But within a few years of the conquest of Peru, the *Biblia Polyglotta* expressed the idea that the Americas had been populated after the flood, by the descendants of Shem, the son of Noah.

Despite the universality of the flood myth, it has almost always been dismissed by scientists as either nonexistent or a local event at best. Giorgio de Santillana, a former professor of the history and philosophy of science at MIT, and Hertha von Dechend, who holds approximately the same post at the University of Frankfurt in Germany, explore the origins of human knowledge in the preliterate world in their classic study of mythology *Hamlet's Mill*. "Myth," they say, "was a language for the perpetuation of a vast and complex body of astronomical knowledge."

Having stated their case, they then go on to force the myths of the great flood into an astronomical abstraction. "There are many events, described with appropriate terrestrial imagery, that do not, however, happen on earth. . . . In tradition, not one but three floods are registered, one being the Biblical flood, equivalents of which are mentioned in Sumerian and Babylonian annals. The efforts of pious archaeologists to connect the Biblical narrative with geophysical events are highly conjectural. There have been floods in Mesopotamia causing grievous loss of life. There still are in the river plains of China and elsewhere, but none of the total nature that the Bible describes.

"There are tales, too, of cataclysmic deluges throughout the great continental masses, in Asia and America, *told by peoples who have never seen the sea, or lakes, or great rivers.* The floods the Greeks described, like the flood of *Deucalion,* are as 'mythical' as the narrative of Genesis. Greece is not submersible, unless by tsunamis.

"The 'floods' refer to an old astronomical image, based on an abstract geometry. That this is not an 'easy picture' is not to be wondered at, considering the objective difficulty of the science of astronomy."

The astronomical explanation then offered by Santillana and Dechend calls upon the constellations that mark the equinoxes. "A constellation that ceases to mark the autumnal equinox," they point out, "gliding below the equator, is drowned."

But would such astronomic views have been enshrined as myth throughout the peoples of the world? The supernova we call the Crab nebula was recorded by the Chinese in 1054. It was also inscribed in a rock painting by Indians on the west coast of America. "The event, revealing the workings of the natural order of the universe," notes Vine Deloria, "was completely ignored in Western Europe, where Christian theologians refused to admit its possibility."

If the Christian astronomers of the time, in search of a constant and predictable order for the heavens, could ignore such a momentous cosmic catastrophe as an exploding star that lighted the skies like another sun, how could the mere movement of the constellations—familiar movements at that—give rise to the worldwide flood myths? The answer is it could not. Worldwide flood myths arise when catastrophic floods wash over a good part of the earth. And there is a great deal of evidence that this is precisely what happened. The proof has not been long recognized, and many false trails have been followed in search of "scientific evidence" of the flood.

Among the first to pick up the scent was an Italian naturalist named Ristoro d'Arezzo, who in 1282 published a treatise that outlined the structure of the world. In it he claimed that the biblical deluge had left clearly visible traces on the land. "And, in fact, as proof that all was once covered by water," wrote d'Arezzo, "near a high mountain peak we dug up many bones of fishes, as well as forms which we would prefer to call snails, but which other people might call sea shells. On one mountain we found colored sand, mixed with large and many small rounded stones whose shape betrayed their provenance in water; a further indication that such mountains were overwhelmed by the Flood."

The fish fossils found by d'Arezzo were dismissed as evidence of the flood by that universal genius Leonardo da Vinci. He reasoned that the deluge could not have carried the fossils hundreds of miles from the sea. "That cannot be," he wrote in his secret mirror writing, "for the Deluge was caused by torrential rainfall which the rivers naturally carried to the sea, along with the dead things that had been washed into them. The rainfall did not draw the dead things from the shores of the sea to the mountains."

Leonardo's conclusion: The fossils were there because the fish that had deposited their bones on those mountains had once lived there. What we now know of plate tectonics seems to support Leonardo's conclusions.

But the evidence of the flood is persistent; it will not go away. In the British Museum are the fragmentary remains of the thousands of clay tablets which once filled the temple library of King Ashurbanipal of Assyria. In 1853 they were uncovered by British archaeologists excavating the ancient city of Nineveh. They were brought to London, where they lay for another twenty years before their true significance was realized.

Then a banknote engraver turned scholar, named George Smith, was given a cardboard box containing a single clay fragment. Smith looked at the ancient cuneiform writing, which looked like the tracks a hen might leave walking across soft clay, and suddenly realized he was reading an account of the great flood. "The first man to read that after more than two thousand years of oblivion," as he put it, laid the tablet down with great solemnity. Then, according to the official British Museum version, "he jumped up and rushed about the room in a great state of excitement and to the astonishment of those present began to undress himself."

Smith's outlandish behavior belies his scholarship, which swiftly reasserted itself, and a few months later he delivered a paper describing his findings of the Babylonian account of the flood. This led the London *Daily Telegraph* to underwrite Smith's journey to Nineveh to search for still more tablets. They added more detail to the story, but another fifty years were to pass before geological evidence of the actual flood was to be uncovered.

In 1923 Sir Leonard Woolley led an Anglo-American expedition to the base of a huge red mound in the heart of the Mesopotamian desert. Within it lies the mutilated stump of what was once the great tower of Ur. Four thousand years ago the tower was a vast square of brick, 120 feet long on each side, 75 feet high, and topped with a roof of gold. It was here that Woolley and his companions began to dig. Six years later they finished, having unearthed the remains of the fabled city of Ur of the Chaldees, along with geologic evidence of the flood mentioned in the ancient Babylonian tablets deciphered by George Smith.

The Woolley group had dug deep into the desert, each layer peeling back evidence of earlier and earlier civilizations, back until they had reached 5000 years into the past, to the height of the ancient and mysterious Sumerian civilization. And still the archaeologists dug until one day the baskets brought to the surface contained nothing but clay, the sort that could have been deposited only by water.

At first, Woolley thought the clay had been left by the Euphrates River, which ran past the gates of Ur 4000 years ago. He soon reconsidered. "I saw," he noted in his diary, "that we were much too high up. It was most unlikely that the island on which the first settlement was built stood up so far out of the marsh."

And so the men kept digging. Deeper and deeper the Arab workers dug, three feet into the clay, five feet, almost ten feet deep, and the wicker baskets came to the surface filled with nothing but clay. As each basket rose to the surface, the men expected to see virgin soil, but then, after ten feet of clay, there came a basket filled with potsherds and the detritus of human habitation. It was incredible.

Beneath the ten-foot layer of clay they had found evidence of yet another human habitation. Moreover, it was an incredibly ancient site, for the pottery was very different from that found above the clay level. Above, the vessels were the product of a sophisticated technology, thrown on a wheel. Below the clay the bowls and jars had been shaped by hand. Nor did the below-the-clay baskets turn up any evidence of metal. The only tools found amid the potsherds were made of flint. The only other finds were the shells of tiny marine creatures. Buried with the remains of human occupation, they could have been left behind in the mud by only one agent—the flood.

Woolley collected his dazzled wits long enough to fire off a telegram to London: WE HAVE FOUND THE FLOOD.

Scholars and scientists were willing to accept Woolley's find as evidence of *a* flood, not *the* Flood. "The legend, born of that long-ago flood, might never have wandered very much farther from its source were it not for the fact that it became a part of the Scriptures, and thus in later ages was zealously carried to every corner of the world by Christian missionaries, offen to become merged with pre-existing traditions indigenous to their localities. Flood traditions are nearly universal, partly because of the efforts of these missionaries but mainly because floods *in the plural* are the most nearly universal of all geologic catastrophes," writes Dorothy B. Vitaliano in her book *Legends of the Earth,* summing up the generally accepted view.

That view must now be changed, for in a pair of olive-gray, silty clay cores dredged up from beneath the Soto Canyon, which cuts through the floor of the Gulf of Mexico off the western coast of Florida, there is evidence of a universal flood. The cores contain within their sediments the skeletons of tiny marine creatures called foraminifera. Radiocarbon dating determined that these creatures died between 11,000 and 12,000 years ago. The skeletons also told Dr. Cesare Emiliani, a professor of geology at the University of Miami, that very suddenly the waters of the Gulf of Mexico had received a massive infusion of fresh water. So great, in fact, was this infusion that it raised the level of all of the world's oceans by 131 feet, a rise so great it flooded all of the earth's coastal regions.

This catastrophic flood, which probably killed most of the earth's population, was caused by melting glaciers. As Professor Emiliani reconstructs events, much of the earth was then covered by glaciers with only the tropical and coastal areas of the temperate zones habitable. Then the climate began to warm up, and the glaciers to melt. A layer of slush formed under the great ice sheets, which served as rollers on which the glaciers surged ahead, and increased their melting, pouring water into the seas.

The Valders Advance, as the southward surge of the North American glaciers is known, was not the only movement of ice sheets. According to Dr. Willard Libby, who received the Nobel Prize for developing his method of radiocarbon dating of biological samples, there is evidence that ice was advancing around the world. Radiocarbon-dated samples from Europe, South America and New Zealand all show movement at the same time as the Valders Advance. The melt waters from all these glaciers caused a tremendous rise in the world's oceans.

Limiting his conclusions to the evidence in the Gulf of Mexico cores, Dr. Emiliani dated the event at 11,600 years ago.

"The time of 11,600 years B.P.," he wrote in the September 26, 1975, issue of *Science,* "when the influx of Laurentide ice melt water into the Gulf of Mexico was highest, coincides in age with the Valders Readvance. This 150 km readvance was, therefore, a surge which led to strong ablation and the observed high concentration of ice melt water in the Gulf of Mexico. The concomitant, accelerated rise in sea level, of the order of decimeters per year, must have caused widespread flooding of low lying areas, many of which were inhabited by man. We submit that this event, in spite of its great antiquity in cultural terms, could be an explanation for the deluge stories common to many Eurasian, Australasian and American traditions."

The popular view of the flood story spreading around the world from a single original catastrophe in Mesopotamia, according to Dr. Emiliani, seem improbable. The Navajo Indians, for example, he points out, regarded the Grand Canyon as the product of a deluge long before their first contact with Europeans. He similarly disposes of the idea that sooner or later every civilization suffers from a great flood upon which it then bases its catastrophe myth. The only logical explanation, he believes, is the massive rise in global sea level that occurred between 11,000 and 12,000 years ago.

The myths of the deluge and other disasters are not ostensible history, but actual eyewitness accounts of catastrophe. They are supported by unassailable scientific evidence. When tested, they have been found true; when questioned, they have had answers.

The ruins of Troy were discovered because Heinrich Schliemann believed that Homer had recorded history and not simply composed epic poetry about heroes and gods who had never existed. Plato in *Timaeus* and *Critias* wrote of the Lost City of Atlantis, destroyed by the anger of the gods. There is now increasing evidence that Atlantis was an island city in the Aegean. In Plato's time it was called Thera, today it is known as Santorini, and it is but the broken arc of what was once a huge circular island. In the year 1500 B.C. Santorini-Thera-Atlantis was torn apart by an incredible volcanic explosion. A chunk of that volcano lies a few miles across the Santorini harbor and still smolders after more recent and less violent eruptions. But there is more, for beneath the 250 meters of ash and pumice upon which the modern buildings of Santorini are built there was recently discovered an ancient city. It matches, in startling detail, the city described by Plato and known to the world as Atlantis.

Not all of the world's myths have had as dramatic a confirmation as Atlantis, the flood, or Exodus, and not all the myths are true. But most of those that tell of great destruction and catastrophe are not tales made up to scare children. They are vivid cultural memories of cataclysmic events that actually happened. Of course, many have been overlaid with the patina of hyperbole and peopled with romantic heroes, but they nonetheless contain basically true accounts of the history of the earth. And that history is often a record of catastrophe.

# 5

# THE DOOMSDAY PEOPLE

Even the most devout atheist is forced to admit that the Bible at the very least is a collection of rattling good stories. It has, of course, also been the inspiration for the moral and theological values of the Western world, as well as for some of its more immoral excesses. But it is to the first book of the Old Testament, Genesis, that a group of eighteenth- and nineteenth-century scientists looked for the inspiration that would explain the history of the earth itself. Unfortunately they were locked into a time frame of approximately 6000 years, a span not originally mentioned in any biblical account, but a bit of marginalia that had become gospel. Held in this minuscule time frame, the geologists of the nineteenth century could not account for the varied appearance of that part of the earth they could see except as the product of cataclysmic events. The view was framed most succinctly by the English geologist and theologian William Buckland, who described a world created in a series of massive upheavals. The final catastrophe, he wrote in 1823, was "a general flood which swept away the quadrupeds from the continents, tore up the solid strata and reduced the surface to a state of ruin." All of the earth's formations and properties, therefore, were explained as "the direct agency of Creative Interference."

William Whewell, a colleague of Buckland's, further refined the dictum by explaining the history of life by the "creative power transcending the operation of known laws of nature."

The ultimate power was usually expressed in the form of the Noachian deluge. The flood was the force used to explain the existence of the confusing objects being found with increasing regularity during the eighteenth century. Engineers digging canals found them, miners working beneath the earth found them, and farmers kept turning them up. These were fossils, the petrified remains of living things. They had been a curiosity since the sixth century B.C., when the ancient Greeks had puzzled over their origin and nature. Eventually they concluded that these were the petrified remains of seashells and other marine creatures. To Empedocles, Pausanias, Herodotus and other Greek thinkers, there was only one explanation: The areas where the fossils had been found must once have been underwater. Anaximander of Miletus, who explained the changing position of the stars with the revolutionary idea that the surface of the earth must be curved, offered the even more startling idea that the fossils represented fish that had died long before and they must therefore be the oldest representatives of the animal world and thus be the ancestors of the entire animal kingdom, including man.

The ideas of the Greeks, however, were lost to the Western world during the Middle Ages, and the fossils that were found were no longer thought to be the remains of once-living creatures, but practical jokes on the part of nature—fanciful creations modeled in stone to look like actually existing animals and plants. They were formed by a force known as *vis plastica* and *virtus formativa*—a magical modeling power probably found in the rays of the sun and the stars. The idea survived until the thirteenth century, when once again the biblical deluge was called forth as the formative agent in the creation of the earth's structure. The year was 1282, and the catastrophic evidence of the flood and its effects on the earth were clearly visible to the Italian naturalist named Ristoro d'Arezzo. D'Arezzo held to the Aristotelian view that fossils were formed by the forces of *vis plastica.* But the more he pondered the matter, the more he came to realize that Aristotle was wrong. These jokes of nature were in actuality the petrified remains of creatures that had once been alive, and their presence was clear evidence that the earth had once been covered by the waters of the flood.

Like tidal floods, the idea rose and fell for the next few centuries as the church first denied that the earth could ever have changed, since it was God's creation and thus perfect to begin with, or that whatever changes had occurred were the product of biblical catastrophes such as the flood.

The literal truth of biblical catastrophes was given additional, chilling evidence toward the end of the seventeenth century.

Doomsday hung like a cloud over all Europe as Halley's Comet made its appearance in the heavens and drew closer and closer. Under its influence, a number of scientists looked to biblical catastrophe to explain the formation and structure of the earth.

For this was an age of inquiry, and such objects as fossils and the curious stratified nature of the earth's crust that were being observed increasingly by miners and engineers demanded some sort of answer. To the church, all the answers were to be found in the Bible, and even less than pious scholars and scientists who sought answers outside the biblical framework were drawn by the very nature of their observations and knowledge back to the biblical accounts of the deluge.

Among the most important of those secular scientists to argue the flood theory was a member of London's exclusive Royal Society—a professor of physics at Gresham College named John Woodward. Looking at the earth, Woodward believed he saw the effects of a universal deluge. A cautious, prudent man, he looked extensively in his efforts to learn as much as he could about the "entire mineral kingdom." Woodward traveled all over England to wherever he had heard of an unusual mineral formation, cave or grotto. Wherever men were mining or wells being dug, Woodward eventually would turn up, observe everything in great detail and finally, after taking copious notes, return home.

"Woodward's notes reveal an almost pedantically exact investigator, certainly not a dilettante visionary," wrote Herbert Wendt, in his excellent history of paleontology *Before the Deluge*. "With equal deliberation he assembled all the evidence that challenged the notion of *vis plastica* and supported the theory of the organic origin of fossils.

First he turned his attention to petrified marine animals. The marine shells looked exactly like their present-day relations on the shores of the seas, resembling the latter in shape, size, material, texture, sculpture, in the composition of the lamellae, the imprints of the muscle strands and other respects. Moreover, the fossil shells showed one common characteristic of marine shellfish: smaller shellfish growing on the shells of larger ones. Thus, Woodward concluded, they must actually be the remains of animals that had once lived, and not merely stones."

Woodward also knew of the evidence of other fossilized creatures, giant mineralized bones that had been dubbed dragons and unicorns. These had already been classed by the German philosopher Gottfried Leibniz as the "bones of sea monsters from an unknown world which were transported by the might of the waves from the ocean to the land." This "might of the waves" was for Woodward a term synonymous with the deluge.

Gathering all the evidence as he saw it, Woodward published his *Essay Toward a Natural History of the Earth* in 1695. In it he claimed that the biblical flood had totally leveled the strata of the earth. As the waters receded, they tore and gouged the earth with incredible violence, thrusting huge masses of earth and rock upward, slashing great gaping wounds elsewhere. Thus, the catastrophic violence of the receding waters created the mountains and valleys we know today and layered the earth in the strata we see. The animals trapped in the flood were bashed and battered, hammered into the very stones in which they were later found as fossils. Thus, the fossils were the remains of animals that had lived before the flood and were identical in every respect to those that are alive today. Those fossils that bore little or no resemblance to extant animals were, according to Woodward, the remains of creatures that are alive today but live at the very depths of the seas, far from the eyes of man.

Other versions of the flood as the essential agent in creating the present structure of the earth were also put forward by other scholars and scientists of the time. Leibniz offered a picture of a world built over huge water-filled hollows. He needed some mechanism other than a rain of merely forty days and nights to produce the sort of flood needed to cover the earth. And so he called upon mighty subsidences to squeeze water out of the hollows

大分
七転八起
姫だるま
魔除けの神
木でこ
大分柞原神社
一文人形

and up onto the surface. "The waters squeezed out of the cavities," Leibniz wrote, "then flooded the highest mountains until they found new access to Tartarus, and after shattering the barriers of the hitherto locked interior of the earth, revealed anew what we today see as dry land."

On that land so many geological facts were being reported by the naturalists of Europe that they were beginning to overwhelm the available theories, and scientists everywhere were desperately searching for a single, unifying concept that would embrace all the evidence. Where, in fact, did fossils fit within the scheme of the earth? And what about the many and different types of rocks? How had they been created? And what force had deposited them in the many-layered structures they formed? Erosion, volcanic eruptions and, yes, the flood—all were forces that had to be fitted within a scheme of the earth that would not only make sense of what the scientists were seeing, but also fit within the theology of the time.

To unify all this new knowledge, a theory later to be called catastrophism was developed. It was largely the product of empiricism; men working in and on the earth—miners, engineers, farmers—looked at the evidence beneath their hands and feet and drew conclusions from it. The leader of these empirical geologists was a professor of mineralogy named Abraham Gottlob Werner, who taught at the ancient and famous mining school at Freiberg, in Saxony. Werner was a brilliant teacher, whose students hung on his every word and then, after graduation, went out into the world praising his name and preaching his doctrine. And Werner's doctrine went beyond a single universal flood. He believed that all of the earth's rocks had been precipitated out of the primal oceans that had originally covered the earth. The first action, he maintained, was chemical, when slates and granites crystallized out; then, as the waters lowered, mechanical forces created chalks and limestones, and, finally, silting, while great torrents lashed the earth and laid bare the mountains and the continents.

Once the land was naked, volcanoes, ignited by coal deposits, erupted and, in their explosive fires, twisted the surface, here and there pushing some rock strata out of the horizontal. For the most part, however, Werner saw the earth girdled in uniform layers of rock that had been left by the waters of the past.

Werner and his disciples were known as neptunists, but their view of the earth was based on a mineralogical rather than on a structural basis. Werner, in many ways a stereotype of the rigid, unbending Prussian, recognized this flaw in his theory and set out to modify it, on the basis of observations in his native Saxony, which he took to represent the entire world. Despite his narrow view of the world, he still managed to devise the first geologic classification system.

"The earth," he declared, "is a child of time and has been built up gradually." Werner made this statement in 1787, and with it, he had gone farther than any other geologist until then by arranging the earth's crusts into layers and formations. Every formation, he told his students, consisted of several strata lying one above the other. These strata lay within five major formations, and each of these formations corresponded to a different age in the earth's history. Werner's system of classification became the basis for all subsequent systems of geologic classification, but it contained a basic paradox. The implication of Werner's system was that the earth was quite old, that long periods of time had been required to build the different strata. But the accepted age of the earth was still, in Werner's time, scarcely advanced from the 6004 years assigned to it by James Ussher.

This was a key question for the naturalists of the day, for that brief period of time was a major barrier to shaping an intelligent concept of the earth's structure and formation.

Among the many eighteenth-century scientists to attack the question was Georges Louis Leclerc, Comte de Buffon, who as the son of a wealthy French family had both the money and the time to indulge his curiosity. He studied law and medicine and traveled throughout Europe, something few scientists of the time had done. On a visit to England he was so impressed by Newton's new mathematics of calculus that he translated the great scientist's work into French. After his election to the Academy of Sciences, Buffon was appointed keeper of the Jardin du Roi, the Royal Botanic Gardens, and was soon fascinated by natural history. Buffon, "who could write only when caparisoned in all the elegance of silk, lace, and braided wig," according to one historian, became the Isaac Asimov of his time.

Over a period of fifty-five years he wrote and published fifty volumes called *Natural History*. They were written for the layman and were immensely popular, for they attempted to find and explain the grand designs that Buffon was convinced were present in nature.

Buffon's vision was of a cosmic catastrophe—the collision of a massive comet with the sun—that gave birth to the earth. Such a catastrophe would have left the earth smoldering and hot for thousands of years, far too hot for the maintenance of life. The cooling-down period before life could arise would have had to be somewhat longer than the paltry few thousand years assigned by Archbishop Ussher to the age of the earth. Buffon was a cautious man, and before becoming the first man in Christian Europe to question the marginal note found in most Bibles that assigned the birthdate of the earth as October 23, 4004 B.C., he decided to explore the subject further.

In the hallowed tradition of modern science, Buffon devised an experiment to prove his theory. He took two metal spheres and heated them until they were red hot and then measured the time it took them to cool. By simple extrapolation, Buffon then calculated the age of the earth at 74,832 years. Life, he believed, might have begun after about 30,000 years of cooling. Buffon put these shibboleth-shattering ideas into a book called *Theory of the Earth*.

Buffon's ability to challenge Archbishop Ussher's date with apparent impunity and avoid the censorship and censure of the church was achieved more by the force of his personality than by any theological trappings he might have hung on his writings. For Buffon was a dazzling figure in the Paris of Louis XV. His courtier's tongue, dripping compliments and scintillating conversation in his wake, combined with his peacock dress to blind both church and secular censors to the ideas in his book and allowed him to offer conjectures that went far beyond accepted creation theology.

Fossils, he believed, represented in many instances the remains of creatures that no longer existed. "It may well be," he said, "that a number of extinct species are among them. Shell fish, ammonites, and the strange fossil bones that have been found in Siberia, Canada, Ireland, and many other regions, seem to confirm this conjecture. For in our times we know of no animal to which such bones could be ascribed. Some are of extraordinary length and thickness. . . . Everything seems to suggest that they represent vanished forms, animals that once existed and today no longer exist."

Now Buffon was really in high gear, reaching toward a concept that would unify the age of the earth, the existence of fossils and biblical catastrophe. For he had come to the conclusion that the earth had gone through successive stages during its cooling period and that each of the different periods had produced many different animals and plants, most of which had been wiped out as a catastrophe brought their particular age to a close. Thus, there had been not one great flood but many, and the structure of the earth and its relics were obvious evidence of flood and other catastrophes.

Much of that evidence was being turned up in the gypsum of Montmartre. As the French workers dug foundations into the Parisian soil, they came up with the fossil bones and teeth of animals long since extinct. All the finds were turned over to a young anatomist at the Jardin des Plantes, Georges Cuvier. Cuvier became to European science what Napoleon was to its politics: its virtual dictator. Biology, zoology, anatomy, geology and most other naturalist sciences were steered along the courses he set. For Cuvier was a brilliant scientist, the apotheosis of Napoleonic science—mechanistic and strictly deductive. True to the Newtonian dictum *Hypotheses non fingo* ("I frame no hypotheses"), Cuvier sought only the evidence of his eyes in determining truth. The Newtonian concept allowed him to develop his theory of correlation, which is the bedrock foundation of comparative anatomy—a science he created.

Cuvier's law of correlation states that the various parts of an organism do not exist alongside one another with no relationship to each other. Rather, they are interrelated. Every living organism forms a complete entity, and no part can change without changing the other parts. Thus, the shape of missing parts can be deduced from the separate parts.

This brilliant piece of logic, based on the deductive reasoning that was then considered the only acceptable approach to science, is today an obvious truism, but at the turn of the nineteenth century it was revolutionary. And it came from simple observations that Cuvier had made while tutoring children in Normandy while he himself was still in his teens. Animals with hooves and horns were herbivores and always, without exception, he noted, had the teeth of grass eaters. Animals with claws had the teeth of predators. Reptiles with fangs were carnivores, while those with solid rows of teeth were vegetarians.

After promulgating his law of correlation and seeing that it held true for every creature in the modern world, Cuvier extended it to fossils. "First seek to obtain the teeth of the fossil animal," he declared. "In most cases you will be able to recognize simply by the molars whether the vanished creature was an herbivorous or carnivorous animal."

It was an incredible statement to make, for most vertebrate fossils are found in bits and pieces, looking more like potsherds than bones, pieces of a jigsaw no one had completed. In fact, no one before Cuvier had ever assembled fossil fragments into a complete animal. The bits of bones that had been collected for centuries were simply isolated representatives of creatures whose original form and functions could be only guessed at.

Cuvier startled first Paris and then Europe with his descriptions of prehistoric creatures based on the fragments of fossils turned up in the Seine basin and the gypsum of Montmartre. Sorting out the teeth, skull fragments and pieces of feet, Cuvier was able to reconstruct two distinct animals: a tapirlike *perissodactyl* with three toes and a cloven-hoofed, hornless, deerlike creature called an *artiodactyl*. Both were extinct creatures from the Tertiary period. Soon after Cuvier identified the remains and described their probable appearance, complete skeletons of both animals were found. They conformed in every detail to Cuvier's reconstruction.

He went on to populate ancient France with giant reptiles, flying lizards and hairy elephants. He became the personification of science with his dramatic demonstrations and attracted the plaudits of his peers and the world at large. Honoré de Balzac, a young writer at the time, was so smitten by one of Cuvier's remarkable reconstructions of prehistoric animals that he virtually deified him in this passage from *The Wild Ass's Skin:*

"Is not Cuvier the great poet of our era? Byron has given admirable expression to certain moral conflicts, but our immortal naturalist has reconstructed past worlds from a few bleached bones; has rebuilt cities, like Cadmus, with monster's teeth; has animated forests with all the secrets of zoology gleaned from a piece of coal; has discovered a giant population from the footprints of a mammoth. These forms stand erect, grow large and fill regions commensurate with their giant size. He treats figures like a poet; when he sets a zero beside a seven it produces awe. He can call up nothingness before you without the phrases of a charlatan. He searches a lump of gypsum, finds an impression in it, says to you, 'Behold!' All at once marble takes an animal shape, the dead come to life, the history of the world is laid open before you."

But that history had to conform to the rigid, mechanistic science that Cuvier had helped create. So certain was he of the truth of his deductive method that he clung to it under the most extreme circumstances.

One night, long after he had become famous, a group of students burst into his room where he lay sleeping. "Cuvier, Cuvier," intoned one student, dressed in a devil's costume, "I have come to eat you."

Cuvier, although startled into wakefulness, replied, "All creatures with horns and hooves are herbivores. You can't eat me." He then went back to sleep.

Cuvier utilized this same mechanistic reasoning to account for the sequential populations of extinct animals. Each species was, in his view, fixed and unalterable. Just as Newton had created a clockwork universe, so Cuvier created a static order of species, each relegated to a time and place, each wiped out in turn by a great catastrophe. Floods, volcanoes, earthquakes, all on a worldwide scale, constantly destroyed the land and reshaped it, moving seabed to mountaintop, dry land to ocean bottom. After each cataclysm, or "frightful occurrence," as Cuvier dubbed it, new species would take the place of those that had been destroyed. "Living things without number were swept out of existence by catastrophes," he wrote. "Those inhabiting the dry lands were engulfed by deluges, others whose home was in the waters perished when the sea bottom suddenly became dry land; whole races were extinguished leaving mere traces of their existence, which are now difficult of recognition, even by the naturalist. The evidences of those great and terrible events are everywhere to be clearly seen by anyone who knows how to read the record of the rocks."

Across the English Channel, however, there were those who read that record of the rocks differently. A few years before Cuvier reached the pinnacle of his prominence a Scottish gentleman-farmer of many parts, a jurist, physician, chemist and physicist, named James Hutton, was also reading the record of the rocks and coming to a totally different conclusion. Quietly and methodically Hutton probed the wild glens and coasts of his native Scotland, seeking a coherent thread, some generalized concept that would explain or coordinate the marked differences in geological structure and formation.

To Hutton the deductive methods of his time were inappropriate to the task, for they ignored the evidence of the present which was available and obvious to anyone who cared to explore it. He chose "to examine the construction of the present earth, in order to understand the natural operations of time past; to acquire principles by which we may conclude with regard to the future course of things or judge of those operations by which a world so wisely ordered, goes into decay; and to learn by what means such a decayed world may be renovated, or the waste of habitable land be repaired."

Hutton in his classic *Theory of the Earth,* first read to the Royal Society of Edinburgh in 1785 and then published in three volumes in 1795, laid down one of the sacrosanct rules of modern geology, the idea that "the present is the key to the past." Thus, Hutton ruled out the catastrophic concept as a means of geologic formation. Floods, convulsive earthquakes and other catastrophes had no role in Hutton's theory of the earth. Rather, Hutton believed the earth had evolved in a rather uniform fashion over long periods of time. "No powers are to be employed," he said, "that are not natural to the globe, no actions to be admitted to except those of which we know the principle, and no extraordinary events to be alleged in order to explain a common appearance."

Thus, time, the great stumbling block of the catastrophists, was available in Hutton's concept in vast, abundant quantities.

On a visit to the Berwickshire coast with his friend mathematician John Playfair, Hutton described his concept of a geologic time frame that stretched back toward infinity.

"The mind seemed to grow Giddy by looking so far into the abyss of time," wrote Playfair of the expedition, "and while we listened to the philosopher [Hutton] who was now unfolding to us the order and series of these wonderful events, we became sensible how much further reason may sometimes go than imagination can venture to follow."

How else to explain Hutton's staggering concept of an earth that had always been and would ever be. "We find no vestige of a beginning—no prospect of an end," he wrote and thus ran squarely afoul of the church. By ruling out catastrophes and declaring that the earth was far older than its 4004 B.C. date of origin as decreed by Archbishop Ussher, Hutton, a pious Quaker, had virtually damned himself. But it was his idea that fire, not flood, had formed the rocks in the earth's crust that led to an article in the Royal Irish Academy charging him with atheism.

Hutton had agreed that some rocks had been laid down as sediment and compressed, but unlike Werner and the neptunists, he did not believe the agency had been the universal deluge. Moreover, he pointed out that other rocks had once been molten in the earth's interior and then brought to the surface by volcanic action. This was the basis for an opposing school, the plutonists, to form and account for, as the neptunists could not, bent and tilted strata and the obvious evidence of worldwide volcanic activity.

"We may perceive the actual existence of those productive causes," wrote Hutton, "which are now laying the foundation of land in the unfathomable regions of the sea, and which will, in time, give birth to future continents."

In Hutton's view the earth was a renewable engine, decaying over the course of time owing to wind, rain and other erosive forces, only to be renewed and rebuilt by the work of great fiery forces beneath the surface of the earth. It was to Hutton "a reproductive operation, by which a ruined constitution [was] again repaired. This decaying nature of the solid earth is the very perfection of its constitution as a living world."

"James Hutton, in other words," wrote Loren Eiseley, the gifted writer and anthropologist, "was the creator of a self-renewing world-machine whose laws of operation were as unswerving as the cosmic engine of the astronomers. In this respect he was following the scientific bent of his time. His misfortune lay in the fact that what had become acceptable in the heavens was still a heresy upon earth."

The accusations of atheism that followed in the wake of his ideas so affected Hutton that he became very ill and died soon thereafter, without being able to refine and clarify his ideas to the point where they were fully to replace the catastrophic doctrine so dramatically espoused by Cuvier and others at the beginning of the nineteenth century.

Onto this scene came an English lawyer named Charles Lyell.

Charles Lyell was born in 1797, the year James Hutton died. But it was not until he left Oxford and had passed the bar that Lyell even learned of the existence of James Hutton and his revolutionary idea that the present is the key to the past. Lyell, while at Oxford, had been fascinated by geology and had attended numerous lectures given by William Buckland, Oxford's first reader in geology, a confirmed neptunist and the most dynamic spokesman of catastrophism in England. But after several trips to the Continent Lyell's own observations turned him into a plutonist, which led him directly to Hutton's ideas of the uniformity of nature.

"No causes whatever have from the earliest time to which we can look back, to the present, ever acted, but those now acting; and that they never acted with different degrees of energy from which they now exert." Behind this statement of Lyell's was the Huttonian idea that all geological changes in the past occurred in the same fashion as they do today, and with the same force, and that catastrophe as the sole cause of geological change, or even a major cause, was impossible

To prove this idea, Lyell published three massive volumes entitled *Principles of Geology*. In them Lyell offered no new laws, created no new theories and laid down no history of regions not previously studied. Rather, he synthesized the ideas of Hutton and others and produced a polemic designed once and forever to destroy totally the ideas of the catastrophists, "to sink," as he put it, "the diluvialists."

"We hear," he wrote in *Principles of Geology,* "of sudden and violent revolutions of the globe, of the instantaneous elevation of mountain chains, of paroxysms of volcanic energy, declining according to some, and according to others increasing in violence, from the earliest to the latest ages. We are also told of general catastrophes and a succession of deluges, of the alternation of periods of repose and disorder, of the refrigerations of the globe, of the sudden annihilation of whole races of animals and plants, and other hypotheses, in which we see the ancient spirit of speculation revived, and a desire manifested to cut, rather than patiently to untie, the Gordian knot."

Lyell's book, clearly written, heavily documented and well presented, was a best-seller and became quite simply the single most influential book in the history of geology.

It was also, according to Harvard's Stephen Gould, "one of the most brilliant briefs ever published by an advocate. It is a melange of precise documentation, incisive argument, and a few of the 'quiddities, quillets [quibbles] . . . and tricks' that Hamlet ascribed to the profession when he exhumed a lawyer's skull from the graveyard. Lyell relied upon two bits of cunning to establish his uniformitarian view as the only true geology."

Lyell accomplished this, according to Professor Gould, by setting up a straw man to demolish. That straw man was, of course, the catastrophist time frame that dated the birth of the earth at 4004 B.C. Most of the leading catastrophists of the time, although not the average churchgoing nineteenth-century edition of the man in the street, had already rejected that idea. Cuvier, Louis Agassiz and even Roderick Murchison, who had accompanied Lyell on many of his geological field trips, were catastrophists who believed in the antiquity of the earth but thought that most of the formations that could be seen and studied were the product of natural catastrophic forces.

The second "bit of cunning" Gould sees is that Lyell's uniformity was a "hodgepodge of claims. One is a methodological statement that must be accepted by all scientists, catastrophist and uniformitarian alike. Other claims are substantive notions that have since been tested and abandoned. Lyell gave them a common name and pulled a consummate fast one: he tried to slip the substantive claim by with an argument that the methodological proposition had to be accepted, lest 'we see the ancient spirit of speculation revived, and a desire manifested to cut, rather than patiently to untie, the Gordian knot.' "

There are four basic components to Lyell's uniformity. The first is that natural laws are constant—that is, uniform in time and space. According to Gould, "this is not a statement about the world; it is an a priori claim of method that scientists must make in order to proceed with any analysis of the past. If the past is capricious, if God violates natural law at will, then science cannot unravel history. . . . The catastrophists agreed; they, too, sought a natural cause for cataclysms and praised Lyell's basic defense of science against theology."

Lyell's second principle is that processes now operating to form the earth's surface should be used to explain the events of the past. This is another facet of uniformity, in this case that geological processes are uniform throughout time. Here again, Lyell is not voicing universal truth but merely pointing out scientific procedure. Only present processes can be seen directly. The past events which cannot be directly observed are best explained as the product of processes which are still at work. On principle number two, no scientist, neither neptunist nor plutonist, neither uniformitarian nor catastrophist, disagreed. Earth-building processes that could be directly observed were obviously to be preferred in seeking explanations of how the structure and form had been achieved. But where Lyell and the catastrophists disagreed was on the subject of catastrophe itself. Lyell felt that present processes alone could explain all of the earth's history. The catastrophists believed that present processes were the preferred explanations but that some events in the past could be explained only by causes no longer at work or, if still active, proceeding at much slower rates than in the past.

This came into conflict with Lyell's third law that geologic change is always slow. gradual and steady. This uniformity of rate rules out all catastrophic events as forces capable of effecting changes in the earth's structure or of playing a role in its formation. Even after Louis Agassiz had proved his ice age theory, Lyell refused to accept the idea that amounts of ice or their rates of flow ever had been very different from those at that moment.

For Lyell's fourth concept of uniformity was that the earth has been essentially the same since its formation. This uniformity of material conditions has been largely ignored by modern geologists and science historians, for it is largely incorrect.

It is also another point of uniformitarianism upon which the catastrophists could not agree. But the idea that they were totally opposed to Lyell and all his uniformitarian ideas is not true. One of the most brilliant scientists of the time, Louis Agassiz, had been a student of Cuvier and was a confirmed catastrophist. A Swiss who had camped out on the glaciers of the Alps and thus had the evidence of his own eyes to attest to their movement, he had been invited to the United States in 1846 to lecture not about glaciers, but about fossil fish. He remained, became an American citizen and spent the last quarter century of his life at Harvard. There he gave numerous lectures in which he reaffirmed his view of the earth's catastrophic history.

"The first two great classes of rocks, the unstratified and stratified rocks," he told his students, "represent different epochs in the world's physical history: the former mark its revolutions, while the latter chronicle its periods of rest. All mountains and mountain-chains have been upheaved by great convulsions of the globe, which rent asunder the surface of the earth, destroyed the animals and plants living upon it at the time, and were then succeeded by long intervals of repose, when all things returned to their accustomed order, ocean and river deposited fresh beds in uninterrupted succession, the accumulation of materials went on as before, a new set of animals and plants was introduced, and a time of building up and renewing followed the time of destruction."

But Agassiz read and applauded Lyell's work, although pointing to their differences of opinion. Agassiz's actual comments recorded in the margins of his copy of *Principles of Geology* were recently turned up by Stephen Gould while browsing through the stacks of Harvard's library. "Agassiz was America's leading biologist and also her staunchest catastrophist," notes Professor Gould. "And yet his marginalia include an impossible contradiction if we accept the standard account of Lyell's achievement. Agassiz's penciled annotations include all the standard critiques of the catastrophist school. They record, in particular, Agassiz's conviction that the summation of present causes over geologic time cannot account for the magnitude of some past events; a notion of cataclysm, he believes, is still required. Nonetheless, he writes as his final assessment: 'Mr. Lyell's *Principles of Geology* is certainly the most important work that has appeared on the whole of this science since it has merited its name.' "

Where Lyell incurred not merely the disagreement of the catastrophists, but their violent opposition, was in his interpretation of fossils. These mineralized bones and shells, branches and leaves of the ancient past were the instrument that had started the quest for understanding the earth and its earliest inhabitants and had first given rise to the theory of catastrophism. Lyell, who had in every other area utilized only the ideas of others and the evidence of his own eyes, disregarded that evidence in his view of the history of life. He firmly believed, despite all the evidence to the contrary, that mammalian fossils would be found in the earliest fossil-bearing beds. The obvious difference between some fossil species and their modern counterparts he discounted. Rather, he took the total fossil record as the representative of but one part of a "great year"—a grand cycle that will occur again when "the huge iguanodon might reappear in the woods, and the ichthyosaur in the sea, while the pterodactyl might flit again through umbrageous groves of tree ferns."

"The catastrophists," says Stephen Gould, "took the literal view. They saw direction in the history of life, and they believed it. In retrospect, they were right."

Modern geology reflects that position. It is, in fact, a blend of both schools, of Lyell's rigid uniformitarianism and the scientific catastrophism of Cuvier and Agassiz. But what Lyell did by clearly spelling out the dynamic nature of the earth's processes was to strike free the shackles of a church-decreed time frame that was forced to deny the bulk of those processes.

There is a great paradox in this, for the Christian church, which fought so dogmatically to preserve the literal truth of its Bible and thus the faith of its believers, had also provided the mechanism by which that literal truth was to be destroyed. "Although we may recognize the frailties of Christian dogma . . ." notes Loren Eiseley, "we must also observe that in one of those strange permutations of which history yields occasional rare examples, it is the Christian world which finally gave birth in a clear articulate fashion to the experimental method of science itself."

To philosopher Alfred North Whitehead this was the essential paradox of modern science: "the sheer act of faith that the universe possessed order and could be interpreted by rational minds."

In exploring the battle between catastrophism and uniformitarianism, in his book *Darwin's Century,* Loren Eiseley notes that the philosophy of experimental science was not impressive. "It began its discoveries and made use of its method in the faith, not the knowledge, that it was dealing with a rational universe controlled by a Creator who did not act upon whim nor interfere with forces He had set in operation. The experimental method succeeded beyond men's wildest dreams but the faith that brought it into being owes something to the Christian conception of the Nature of God. It is surely one of the curious paradoxes of history that science, which professionally has little to do with faith, owes its origins to an act of faith that the universe can be rationally interpreted and that science today is sustained by that assumption."

It was the translation of faith into scientific method then that was Charles Lyell's greatest accomplishment, for with *Principles of Geology* he created a new science, based on the scientific method. And he did it with existing ideas and observable evidence.

So important was Lyell's contribution to the scientific revolution that upon the centennial of his death on February 22, 1975, the prestigious British journal *New Scientist* declared:

"Twice in the last 500 years acceptance of a major scientific advance has required man to abandon a theological dogma widely held through Christendom. Charles Lyell who died 100 years ago this month played a central part in the second of these revolutions. His thinking linked that of the Scottish geologist James Hutton . . . with that of Darwin. . . ."

Herein may lie Charles Lyell's ultimate contribution to the scientific revolution—his influence on Charles Darwin. Not only was Lyell the scientist Darwin respected most, but his theory of uniformitarianism, which initially disavowed evolution, also provided the time needed for the evolutionary process to work. And it was by reading *Principles of Geology* while sailing aboard the *Beagle* that Charles Darwin, a confirmed creationist, became the creator of the theory of evolution.

# 6

# VOYAGE TO EVOLUTION

In the pantheon of science no one rates higher than Charles Darwin. He is right up there with Newton and Einstein, the author of a scientific revolution. Like his inspiration and occasional mentor, Charles Lyell, Darwin essentially authored no new thought, but he did synthesize a great many existing ideas that had been lying about for years, dusty and unused or vibrant and pulsing, but still unused by the scientists of the time.

"State broadly [that there is] scarcely any novelty in my theory," Darwin wrote in a disclaiming introduction planned for the preface to the book he had not yet written. "The whole object of the book is its proof, its extension, its adaption to classification and affinities between species.

"Seeing what von Buch [Humboldt], G. H. Hilaire and Lamarck have written, I pretend to no originality of idea (though I arrived at them quite independently and have read them since). The line of proof and reducing facts to law [is the] only merit, if merit there be, in the following work."

The assembly of his proof and the reduction of facts to law were to take Darwin twenty years of writing and rewriting. The book might still never have been published had not Alfred Wallace, another naturalist, sent a manuscript to Darwin for his opinion. It contained, although with considerably less documentation, the selfsame ideas Darwin had pored over so laboriously for so many years. The Wallace manuscript goaded Darwin into finally making his ideas public, something his friends, including Charles Lyell, had been pleading for him to do for years. And so the Darwinian revolution began on July 1, 1858, with the meeting of the Linnean Society in London.

According to the *Journal of the Linnean Society,* that meeting heard four reports: a letter from Sir Charles Lyell and Joseph Hooker to the society's secretary, J. J. Bennett; an "Extract from an unpublished work on Species, by C. Darwin, Esq., consisting of a portion of a chapter entitled, 'On the Variations of Organic Beings in a state of Nature; on the Natural Means of Selection; on the Comparison of Domestic Races and the True Species' "; an "Abstract of a letter from C. Darwin, Esq., to Prof. Asa Gray, Boston, U.S. dated Down, September 5th, 1857"; and a paper "On the Tendency of Varieties to depart indefinitely from the Original Type. By Alfred Russel Wallace."

As an eminently proper Victorian gentleman, Darwin, although first to have developed his ideas on evolution, nonetheless sent Wallace's manuscript on to the Linnean Society along with his own extract. A few months later he finally sent to the publisher the book he had started in 1844, the synthesis of everything he had seen and learned on his five-year voyage aboard HMS *Beagle,* which had ended in 1836.

*On the Origin of Species, by Means of Natural Selection, or the Preservation of Favoured Races in the Struggle for Life,* to give the book its full title, was published in November 1859. It not only revolutionized science, but also sent profound shock waves through all intellectual and cultural fields and altered forever the picture we have of man and his place in the universe.

Stated simply, Darwinian evolution rests on five main points:

1. All species produce offspring in excess of the number that can survive.
2. Adult populations in any region tend to remain constant, and therefore, there is an enormous death rate. (Most biologists believe the first part of number 2 is wrong, the second part largely correct.)

3. There must be a struggle for survival which the majority of creatures lose.
4. The competitors vary in many small characteristics, and these will affect the chances for survival.
5. The result of the four previous conditions is that the organism best able to survive the conditions transmits its more adaptive traits to future generations.

In these ideas Darwin had provided the final blow to a theologically dominated science and created a new synthesis that would affect not only science, but all thinking from that moment onward. The neat, steady world of Newtonian physics, fixed and immutable, which had offered the rock-solid basis for the idea of the creation of fixed species, had been shaken to its core. And while many historians of science look on this as the final deathblow to catastrophism, the fact was that many catastrophists had already moved very far from the idea of fixed species and saw a flow and direction to life, albeit a direction that was dictated by catastrophe.

Darwin himself, when he first went aboard the *Beagle* at the tender age of twenty-two, was a catastrophist, cast firmly in the mold of Cuvier and convinced of the immutability of the species.

The captain of the *Beagle,* Robert Fitzroy, was also a catastrophist and believed the voyage would provide the perfect opportunity to substantiate the Book of Genesis.

"As a naturalist," wrote Alan Moorehead in his book *Darwin and the Beagle,* "Darwin might easily find many evidences of the Flood and the first appearance of all created things upon the earth. He could perform a valuable service by interpreting his scientific discoveries in the light of the Bible. Darwin, the young clergyman-to-be, was very ready to agree."

And indeed there were occasions when the incredible features of the South American landscape Darwin was to explore seemed certainly to have been born in the crashing tumult of catastrophe.

Finding a great arched fragment of rock on a Patagonian peak, he asks, "Must we believe that it was fairly pitched up in the air, and thus turned? Or, with more probability that there existed formerly a part of the same range more elevated than the point on which this monument of a great convulsion of nature now lies. . . . If during the earthquake which in 1835 overthrew Concepción in Chile, it was thought wonderful that small bodies should have been pitched a few inches from the ground, what must we say to a movement which has caused fragments many tons in weight, to move onwards like so much sand on a vibrating board, and find their level? I have seen, in the Cordillera of the Andes, the evident marks where stupendous mountains have been broken into pieces like so much thin crust, and the strata thrown on their vertical edges: but never did any scene, like these 'streams of stones,' so forcibly convey to my mind the idea of a convulsion of which in historical records we might in vain seek for any counterpart: yet the progress of knowledge will probably some day give a simple explanation of this phenomenon, as it already has of the so long thought inexplicable transportal of the erratic boulders, which are strewed over the plains of Europe."

Among the books Darwin took on board the *Beagle* just before its departure from England on December 27, 1831, was the first volume of Charles Lyell's just published *Principles of Geology*.

Darwin read Lyell, at first with the caution of his Cambridge mentor the Reverend John Henslow in his ears.

"Read it by all means," wrote Henslow, who had given the book to Darwin, "for it is very interesting, but do not pay any attention to it except in regard to facts, for it is altogether wild as far as theory goes."

But it was precisely the theory that attracted Darwin, for he had begun to question all the traditional ideas of geology he held. Cuvier's catastrophism seemed too rigid, too mechanical to explain the variety of geological phenomena he saw in South America. Here he saw as if through Lyell's eyes the stratification of the rocks, the slow rises and falls of the land, the upthrusting growth of mountains and the effects of the ice age. On every hand there seemed to him to be the evidence of Lyell's primary principle that the earth had been shaped in the past by the same forces that were still at work in the present.

"For Darwin," notes University of New Hampshire's Science Historian Professor Charles J. Schneer, "confronted on the single voyage with a variety of geological phenomena nearly equal in sum to the total that had been previously studied by geologists, Lyell's *Principles* were a method of organization and interpretation. In place of telling Darwin what he would find, the *Principles* were open-ended. Lyell described processes of erosion and deposition as he had seen them and understood them. But these were not offered as the sum of geological processes. In observation, other processes would make themselves manifest. Each phenomenon would contain within itself the trace of its history."

Climbing an Andean peak above Valparaiso, Chile, for example, Darwin was struck by the evidence of the growth of these mountains. At 7000 feet he discovered a small forest of petrified pine trees, surrounded by rocks that had once been lapped by waves. Darwin concluded that once the forest had stood along the shores of the Atlantic Ocean, now 700 miles away. And still higher, at 12,000 feet above sea level, he found a bed of fossil seashells. It was clear to Darwin that the entire Andean mountain chain and the coastline that lay at its feet were once underwater and had been thrust upward in recent geologic time. As the Andes rose, they had been first a chain of forested islands, then a vast mountain range lifted so high the cold killed off the trees and preserved them so that eventually their trunks turned to stone.

It was at this time, July 1834, that the third volume of Lyell's *Principles of Geology* reached Darwin and virtually finished his conversion from a catastrophist to a uniformitarian.

The *Principles* were significant to Darwin in more ways than one. "Without the public revision of attitude on the subject of time and natural forces working over inconceivably long intervals," wrote Loren Eiseley in *Darwin's Century*, "Darwinism would have had little chance of acceptance. Moreover, it is unlikely that without the influence of Lyell's book Darwin would have conceived or put forth his theory."

There is, moreover, the paradox of Lyell's not believing in organic evolution until many years after *Origin of Species* was first published. "One can scarcely resist the observation," says Eiseley, "that the *Origin* could almost literally have been written out of Lyell's book once the guiding motif of natural selection had been conceived."

Darwin was able to see the product of natural selection in the Galápagos Islands. This bleak, desolate group of fourteen islands lies on the equator, in the Pacific Ocean,

600 miles off the coast of Ecuador. Born of catastrophic underwater volcanic eruptions, the islands contain a relatively small variety of animal life, but that little is remarkably odd. Giant sea turtles weighing as much as a ton, some small lizards and a few different species of birds, although of vast numbers, constitute the islands' primary life.

It was among the giant turtles that Darwin first noticed the peculiar fact that was to set him to thinking, for virtually the rest of his life, about the question of evolution.

"I have not as yet noticed by far the most remarkable feature in the natural history of this archipelago," he was later to write of the Galápagos; "it is, that the different islands to a considerable extent are inhabited by a different set of beings. My attention was first called to this fact by the Vice-Governor, Mr. Lawson, declaring that the tortoises differed from the different islands and that he could with certainty tell from which island any one was brought.

"I did not for some time pay sufficient attention to this statement and I had already partially mingled together the collections from two of the islands. I never dreamed that islands, about fifty or sixty miles apart, and most of them in sight of each other, formed of precisely the same rocks, placed under a quite similar climate, rising to a nearly equal height, would have been differently tenanted; but we shall soon see that this is the case. It is the fate of most voyagers, no sooner to discover what is most interesting in any locality, than they are hurried from it; but I ought, perhaps, to be thankful that I obtained sufficient material to establish this most remarkable fact in the distribution of organic beings."

And so it began, the attempt to understand the remarkable distribution of species and from it to develop a theory of evolution that would shake science to its very foundations. The chain of thought the tortoises began in his mind was forged into links of logical strength by his discovery of the finches of the Galápagos. Each island had a slightly different species of the bird, prompting Darwin to characterize them as "a most singular group of finches."

There were fourteen separate species, and not one was identical to the mainland finches or, for that matter, to any other species of finch anywhere else in the world.

Darwin felt that an original pair of seed-eating finches had come from the mainland, and their offspring had spread to all the Galápagos Islands.

The descendants of the original finches gradually evolved into different forms. Some developed the ability to eat seeds of one sort, some of another, while others became insect eaters. For each diet and way of life, the finches had developed a particular beak, a specific size and a particular social organization to take advantage best of what the environment had to offer. The original founding species of finch did not do this on the mainland owing to the pressure of competition from other songbirds. On the Galápagos the finches were the only songbirds and had no competition. They were free to develop in the fashion that would best take advantage of the available resources.

Faced with the fact of evolution, for surely it was inconceivable that a special act of creation had created the fourteen different species of finch specifically for the fourteen islands of the Galápagos, Darwin still had no real inkling of the mechanism by which evolution was accomplished. He had only the evidence of species variation, not its cause, and without the mechanism he was not much farther along than the evolutionists who had preceded him.

The idea of evolution had been kicking around for some time. Buffon had hinted at it some seventy-five years before, and even his grandfather, Erasmus Darwin, had authored a theory of evolution. The first well-thought-out evolutionary theory, however, was the work of the French naturalist Jean Baptiste Lamarck.

"Species pass into one another, from simple infusoria up to man," he wrote in 1809 in *Philosophie zoologique*. "The fossil forms of organic life are the real, genuine forerunners of our present living beings." Not content to have articulated a theory of evolution, Lamarck then sought to describe its mechanism. "In every animal that has not yet passed the peak of its evolution, an organ is gradually strengthened by frequent and continuous use; it is thus developed, enlarged, and acquires greater vigor proportionally to the duration of use. Consistent nonuse of an organ, on the contrary, imperceptibly weakens it, causes it to deteriorate, gradually reduces its abilities and ultimately causes it to disappear."

Lamarck believed that creatures actually made attempts to change their own characteristics in order to benefit themselves. As an example, he offered the giraffe, then a newly discovered, exotic creature to Europeans. Lamarck envisioned it originally as a short-necked deer that kept stretching to reach the leaves of trees, until succeeding generations grew longer and longer necks. This led to his theory that such acquired characteristics could be inherited. But what of the development of the giraffe's protective coloration? Surely it could not have willed itself to develop a blotched skin? Even without this serious flaw in the scheme, it could not have been accepted, for Cuvier, the outstanding naturalist of the time, despised Lamarck and his theory of evolution because it denied the role of catastrophism. And in debate and lecture, Cuvier demolished Lamarck and his theory of evolution.

By 1836, the year Darwin finally returned to England from his round-the-world voyage, Cuvier had been dead for four years, and the time and the world seemed ready for a new theory of evolution. The problem was that Darwin had no theory, only the fact of evolution itself. The mechanism by which it took place eluded him, but he set to work collecting everything he could find on variation of species. And the key began to emerge from the barnyards and farms of England, where British farmers, like those all over Europe, had been selectively breeding animals for centuries.

"I worked on the Baconian principles," Darwin wrote, "and without any theory collected facts on a wholesale scale, more especially with respect to domesticated productions, by printed inquiries, by conversation with skillful breeders and gardeners, and by extensive reading. . . . I soon perceived that selection was the keystone of man's success in making useful races of animals and plants. But how selection could be applied to organisms living in a state of nature remained for some time a mystery to me."

But once again, Darwin, a voracious reader, was to pick and read "for amusement" something that was to have a profound impact upon his thought. And just as Lyell's *Principles of Geology* with its uniformitarian ideas was to help shake him from the catastrophist view, so a vision of catastrophe was to give him the key to evolutionary processes. In October 1838 he read Thomas Malthus' *Essay on the Principle of Population.* Published originally in 1798, it was a vision of the apocalypse that

Malthus predicted would soon overtake the world through overpopulation.

"Throughout the animal and vegetable kingdom nature has scattered the seeds of life abroad with the most profuse and liberal hand," Malthus wrote. "She has been comparatively sparing in the room and nourishment necessary to rear them. The race of plants and the race of animals shrink under this great restrictive law. And the race of man cannot by any effort of reason escape it. Among plants and animals its effects are waste of seed, sickness, and premature death. Among mankind, misery and vice."

Here was Darwin's revelation for the mechanism by which evolution worked. The struggle for existence that Malthus postulated as the product of overpopulation Darwin saw as the means of natural selection. "It at once struck me," he said, "that under these circumstances favourable variations would tend to be preserved and unfavourable ones to be destroyed. The result of this would be the formation of a new species."

Thus, nature would select out, by harshness of an environment, by the ability to escape a predator or catch a prey, only those members of a species that were best fitted to survive. In the case of entire species, changing conditions, such as the draining of the shallow seas or the coming of an ice age, even though such changes might occur over millennia, would wipe out entirely those species that could not adapt to the new conditions.

"As many more individuals of each species are born than can possibly survive," wrote Darwin, "and so consequently there is a frequently recurring struggle for existence, it follows that any being, if it vary however slightly in any manner profitable to itself . . . will have a better chance of surviving and thus be naturally selected."

Before publishing *Origin of Species,* Darwin spent years poring over his vast collection of specimens and notebooks. He spent eight years writing a book on barnacles and by 1842 had sketched out his ideas on natural selection in a brief thirty-five-page outline. But still the idea of natural selection steeped, and Darwin mulled, gathering evidence, amassing documentation, until goaded by the Wallace manuscript, he finally published his revolutionary theory on the origin of species.

The storm that exploded over Darwin's book is well documented and has become part of the mythology of science. Less well known is the fact that even among his staunchest champions, Professor Asa Gray in the United States, Thomas Huxley, who called himself "Darwin's Bulldog," in England and Ernst Haeckel of Germany, none actually accepted the radical idea of natural selection. Confirmed evolutionists, they still could not accept the mechanism Darwin had offered that made evolution possible.

"Darwin rests in Westminster Abbey, near (if at the feet of) the immortal Newton," wrote Harvard's Stephen Jay Gould. "But he lies there because he convinced the world of the fact of evolution, not because his theory of natural selection triumphed in his day. I cannot think of a single unambiguous supporter of natural selection among Darwin's contemporaries."

How could they be expected to accept a mechanism without knowing upon what it would work? And herein lies Darwin's true genius. For the means by which variations occur, and the substance by which they are transmitted from one generation to the next, were not only undiscovered and unknown, but not even remotely suspected.

But two years before Darwin's book was published an Austrian monk named Gregor Johann Mendel began growing sweet peas in the garden of his monastery. When, after eight years, he became too fat to bend over and tend the peas, he put down what he had observed about their growth in a paper that was published in the *Transactions of the Brünn Natural History Society* and promptly forgotten. But in that paper Mendel had outlined the laws of heredity.

He found that specific characteristics, such as height, color and size of blossom, were passed on to the new plant by its parent. Each of the characteristics was governed by something which Mendel called factors and which we now call genes. These genes, he found, were transmitted as complete units and retained their individuality. When a tall pea plant was crossed with a dwarf plant, Mendel discovered, the progeny of the cross were either tall or dwarf, but never in between.

Although Mendel's paper, and a second one, published in 1869, were completely ignored for the rest of the century, they contained the rules by which natural selection could be shown to work undiluted by vast amounts of time. And this was central to Darwin's theory of natural selection, for it required vast amounts of time in which to work. Darwin had been quick to see that in modern animals the modifications from one generation to the next were so slight as to be imperceptible within the lifetime of the observer or even within the memory of the entire race of man. But the evidence of the thousands of fossils he had seen and collected had convinced him that changes had indeed taken place in present-day species, and in fact, these modifications were the proof of evolution. But any substantial change in the body of an animal that would illustrate natural selection would require thousands, even millions of generations to occur, and that would mean that hundreds of millions of years must have passed since the fossils had been laid down. Moreover, he believed that there must have been a very long period of time, further millions of years, before any fossils ever appeared, when the primordial seas had been filled with soft-bodied animals that had simply melted back into the seas without leaving any trace of their ever having lived on earth.

So Darwin postulated an earth that had been spinning about the sun for hundreds of millions of years. This idea was in direct conflict with the pronouncements of William Thomson. Lord Kelvin, a mathematician and physicist and one of the towering figures of British science. Lord Kelvin favored the theory that the earth had broken off from the sun and was therefore originally at the same temperature as the sun. The time required for the earth to cool down to its present temperature Kelvin calculated at 40 million years. This, he declared, was the age of the earth, and even that, he believed, was a generous estimate, for when the earth was younger and hotter, it probably lost heat at an even greater rate than at present.

Darwin ran squarely afoul of Kelvin's limits for evolution, and, of course, the slow weeding and sorting process of natural selection required far more than a mere 40 million years to achieve the species that now inhabited the earth. "I am greatly troubled at the short duration of the world according to Sir Thomas W. Thomson [Kelvin], for I require for theoretical views a very long period before the Cambrian formation," wrote Darwin.

Kelvin, however, used his figures to dismiss Darwin's theories as pure nonsense. "We find at every turn," he wrote, "something to show the utter futility of Darwin's philosophy." Then, in 1893, as if adding a final blow, Lord Kelvin reduced his estimate of the age of the earth to 24 million years.

This problem of time was becoming the greatest stumbling block to Darwin's theory of natural selection. For while most scientists had come around to accepting the idea of evolution, they had not accepted, indeed could not accept, natural selection. Part of the problem was time. If natural selection required such vast amounts of time, the uniformitarians were prepared to give it to Darwin's theory, for they too required vast eons to support their thesis of slow, gradual change as the only mechanism of geological change. But time was paradoxically also a weakness in the theory of natural selection. Darwin saw nature creating variations in each generation of a species and natural selection picking out those variations to perpetuate the good and doom the bad. But nature required a great deal of time to try out these variations over many generations. During this time, if there were unrestricted and random mating among the various members of the population, the good characteristics would be mingled with the bad, giving rise to some wishy-washy trait that was neither good nor bad, capable of neither ensuring survival nor guaranteeing doom.

Mendel's laws of heredity solved that problem, for they demonstrated that specific traits, such as Darwin's good characteristics, would not be mixed and blended into a mishmash of indeterminate traits. They would, if they survived, remain in their original form, passed intact from generation to generation by the pressures of natural selection.

But of course, Darwin did not know about Mendel's work, and Mendel, who had read Darwin and been impressed enough to annotate his copy of *Origin of Species,* never once mentioned Darwin in his own papers.

So while the controversy raged for the rest of the nineteenth century with theologians and scientists ranged on the one side and primarily scientists on the other, the actual laws of heredity, which explained that natural selection could indeed work, remained buried in an obscure volume, unread, unrecognized and unvalued until the year 1900.

Then a man named Hugo De Vries, a Dutch botanist who had read Darwin and been absolutely smitten with the idea of evolution, applied himself to the problem of natural selection. He saw that it contained no explanation for the manner in which characteristics could vary among individuals of a species. De Vries devised a theory that different characteristics might vary independently of each other and recombine in many different combinations, but still retain their individuality.

"Attributes of organisms consist of distinct, separate and independent units. These units can be associated in groups and we find, in allied species, the same units and groups of units. Transitions, such as we so frequently meet with in the external form both of animals and plants, are as completely absent between these units as they are between the molecules of the chemist," wrote De Vries.

He had in essence rediscovered Mendel's laws, and in searching the literature before publishing his own paper, he rediscovered Mendel. When De Vries finally published, he did so with astonishing modesty, announcing that his own work was merely a confirmation of Mendel's discovery.

A few months later two other botanists also published similar, but independently arrived-at, conclusions and also gave credit to Mendel. Such virtue cannot go unrewarded.

De Vries had used the evening primrose as his experimental plant. Among the varieties of primrose he observed additional characteristics that would suddenly appear, characteristics that neither parent had possessed. And this is where De Vries went beyond Mendel and leaped to the idea of mutation—the concept of a catastrophic change in the unit or gene carrying the characteristic. For in the primrose, De Vries thought he was witnessing evolution at work in his own flower bed, experimenting with "a species which has been taken in the very act of producing new forms."

De Vries thought he was witnessing the creation of new forms by the process of mutation. What he could not know was that the changes he saw were brought about by a plant whose genetic machinery was inherently unstable because it had an unequal number of chromosomes. And ultimately his observations of the evening primrose were not the evidence he needed to say "the origin of species is no longer to be considered beyond our experience."

De Vries' ideas, although springing from an abnormal experimental subject, gave a sudden and needed push to natural selection. Paradoxically it also posed a challenge to the Darwinian idea, for in De Vries' eyes, natural selection was not enough by itself to power evolution through its many variations. He felt that mutations which could produce sudden and dramatically large changes were the proper mechanism by which evolution worked. Natural selection, in De Vries' view, was merely a vast net on which the mutations fell. Those that fitted through the mesh were perpetuated; those that did not died.

De Vries' theory caused a fierce reaction among what had become the uniformitarian Darwinists, who believed that natural, gradual selection was everything and that large, sudden mutations within species were meaningless. De Vries responded coldly: "The general belief in slow changes has held back science during half a century."

The dispute was finally resolved by the mathematicians. They were able to show that such things as Mendel's factors, De Vries' mutations and, in fact, all degrees of change and difference in natural populations could be analyzed by mathematics. Among the mutationists' primary proofs was the phenomenon of insect mimicry. Certain moths, for example, normally tasty morsels for birds, carried markings that made them look at first glance more like a type of butterfly that was decidedly distasteful. For the mimicking moth it was a survival trick, an adaptation, said the mutationists, that was proof of mutation and not slow, gradual natural selection of the trait.

But a mathematician named Sir Ronald Aylmer Fisher proved that only natural selection could create such intricate adaptations as the matching of a mimic to a model. The occurrence of both insect patterns and shapes by the random method of mutation was shown to be mathematically impossible.

After a number of other proofs were brought to bear, De Vries' dramatic and sudden means of advancing evolution was reduced from a major factor to a supporting role. Mutation, it was decided, merely produced the raw material for changes and variation. Without mutation life would develop a deadly sameness, and its myriad species would be unprepared for sudden or even slow catastrophic changes in their environments. And even though most mutations were detrimental and rapidly lost when their carriers died long before they were capable of passing the new trait along by reproduction, a few small mutations were always being introduced to provide enough genetic diversity to cope with a new situation should the need ever arise. "The function of mutation," wrote Fisher, "is to maintain the stock of genetic variance at a high level."

Slight changes, such as a small change in bone structure that ultimately enables a fish's fin to become a leg, tend to be preserved within the species. Such changes, induced by mutation, would tend over a period of time to become firmly established throughout the species. "It has often been remarked, and truly, that without mutation evolutionary progress, whatever direction it may take, will ultimately come to a standstill for lack of further possible improvements," said Fisher. But the possibilities for modification and variation were virtually limitless. Fisher calculated that if a species had only 100 characteristics divided between male and female, more than 1,000,000,000,000,000,000,000,000,000,000 different genetic combinations were possible when two of its members produced progeny.

On that basis Fisher declared, "It has not so often been realized how far most species must be from such a state of stagnation, or how easily, with no more than one hundred factors a species may be modified to a condition considerably outside the range of its previous variation."

The genetic variation Fisher had shown was possible was the key to survival. The Darwinian idea of tooth-and-claw survival of the fittest was transmuted by the mathematics of Fisher and others into a vast natural pool well stocked with all the genetic tools needed for survival of the species, available on call, as needed.

Darwinian evolution, De Vries' mutations and Fisher's mathematics were finally put together into one elegant synthesis in 1937 by Columbia University geneticist Theodosius Dobzhansky in his book *Genetics and the Origin of Species.* Variation, according to Dobzhansky, continually arises in all species by means of mutations in genes and chromosomes. These variations are shuffled and passed on by the mechanism of sexual reproduction into new patterns of variability in the offspring. These are then put to the test of natural selection, and those that are fit pass on the most adaptive traits.

Dobzhansky also dealt with one of the more destructive aspects of Darwinian evolution—his concept of the "survival of the fittest," a phrase which he had picked up from English sociologist Herbert Spencer and which he thought was "more accurate" than "natural selection." Survival of the fittest had become the "scientific" cornerstone of the Industrial Revolution, imperialism and social Darwinism, a rationale that sought to explain the excesses of capitalism as the law of the jungle which people had to follow.

"Phrases such as 'struggle for existence,' 'nature red in tooth and claw,' 'eat or be eaten' were very freely used, especially by popular writers on evolution, in the late nineteenth and early twentieth centuries," wrote Dobzhansky. "This phraseology seemed to appeal to the emotions of many people of those days. With the development of genetics and with a change of the intellectual climate during the current century, it began to be realized that the fierceness of the struggle for existence leading to natural selection was greatly exaggerated. It is simply the fit, rather than the 'fittest,' who survive.

"Evolutionary success is determined by the ability of the carriers of a given genotype to transmit their genes to the greatest possible number of individuals in the following generations."

" 'Struggle,' 'competition,' and like expressions have a metaphorical meaning when applied to natural selection which should not be confused with their usage when applied to human affairs," Dobzhansky went on to say. "Thus trees 'struggle' against the danger of being felled by wind by developing stronger root systems; mammals and birds 'struggle' against cold by developing heat insulation, temperature regulation or by remaining dormant during winter months; desert plants 'struggle' against dryness by having leaves transformed into spines. Plants and animals 'compete' for food when food is scarce, but they do not necessarily fight against one another."

Thus, a century and a half after Darwin set foot on the *Beagle,* the controversy over his theory of evolution continues. No longer do the combatants argue over angels and apes as human ancestors, but rather they dispute over the effects of chemical changes on the molecular level.

In the process, science has eliminated the catastrophists and all hint of catastrophe. At the same time we may have missed the essential Darwinian message.

"Never forget," he wrote in *Origin of Species,* "that every single organic being may be said to be striving to the utmost to increase its numbers."

The message was echoed by Bertrand Russell in his *Philosophy:* "Every living thing is a sort of imperialist, seeking to transform as much as possible of its environment into itself and its seed."

It is the Malthusian note of doom that gave Darwin the mechanism of natural selection. Once again that note is being sounded and we refuse to listen. For we are exploiting nature on an unprecedented scale, converting more and more of the resources designed to support a wide spectrum of life into the means of merely producing more human beings.

"Man," notes Harvard biologist George Wald, "is the first living species, animal or plant, on this planet that has ever been threatened by its own reproductive success."

Herein may lie the ultimate catastrophe: a species that could conceive a Darwin, receive his message of evolutionary glory and then miss the point because of an inherent belief that catastrophe is not part of the biological scheme of things.

# 7

# EVOLUTION BY CATASTROPHE

One hundred years after *Origin of Species* was first published, in November 1959, the leading evolutionists in the world met at the University of Chicago. Their centennial subject: "Evolution After Darwin." The single most often mentioned word was a set of initials—DNA. For with the discovery of the structure of the DNA molecule in 1953, science had also discovered the unit of genetic machinery on which natural selection works. Here was the chemical thread that led back to the beginning of life. A thread, the scientists believed, that "appears to depend on self-replicating and self-varying (mutating) strings of DNA, and these self-replicating and self-varying properties inevitably lead to natural selection."

DNA stands for deoxyribonucleic acid. It is a complex chemical molecule composed of sugars and phosphates arranged in a twisted double helix. Each of the two strands is connected by pairs of nucleic acids. The order in which those nucleic acids are arranged on the helix determines the hereditary message it carries and sends on to the cell in whose nucleus it resides. When male and female reproduce, the DNA is recombined to create a new, unique individual whose genetic information has been selected from the variety of possibilities contained in the parents' DNA.

Where nonsexual reproduction is concerned, single-cell organisms such as bacteria or plant and animal body cells simply divide, with one-half of each DNA strand going to each of the two daughter cells. They then reproduce, from materials within the cell, the other half of the DNA strand. Under ordinary circumstances there is no recombination of genetic characteristics, the identical genetic information is passed on to each of the new daughter cells, and they are in almost every respect genetically identical to the cell from which they originally rose.

The system, although essentially very simple, is in its detail extremely complex. A single gene, for example, might contain 2000 nucleic acid bases, arranged in a specific order. Inside a virus, which may consist simply of a DNA coil 1/2000 of an inch long, surrounded by a protein coat, as many as 170,000 nucleic acids may constitute its genetic message. A human cell containing the most detailed of all genetic messages has as many as 6 billion bases in its DNA molecule. The arrangement of the nucleic acid bases constitutes a code which orders the cell in which it resides to construct itself in a certain fashion. The effect of the entire mechanism is not only to produce blond hair or blue eyes, but to construct all the enzymes which power the biochemistry of the body and thus make life possible.

In so complex an information system some errors are inevitable at virtually every step of the way. During the actual act of conception, when the recombination of DNA molecules takes place and when the newly created organism is developing, the genetic materials are most vulnerable to catastrophic error. The catastrophe can take the form of an errant X ray smashing through a cell wall and dislodging a nucleic acid from its position on the chain. Perhaps catastrophe is a chemical agent passing through the placenta and combining with a number of nucleic acids to disrupt the code entirely. Or perhaps the genetic catastrophe is caused by a virus invading the developing cells and scrambling their DNA machinery so badly a wholly new message is received.

Even the process of recombination is subject to a host of potentially catastrophic events. Entire DNA molecules can be bumped from their positions on the gene and thus produce a different message from the one intended. The result of all these catastrophes is what biologists call mutations. Early in the earth's history, while the planet itself was still evolving, the most readily available agent for revolutionary change was catastrophe.

The doomsday threat that overhung every newly created living thing when the earth was young was, in fact, the essential mechanism of natural selection. "Time and again," wrote MIT biologist and Nobel laureate Salvador Luria, "conditions became critical. New inventions, by mutations and recombinations of genetic materials, were the key to survival. Consider the history of the devices by which organisms obtain energy for life. Each of these —fermentation, photosynthesis, respiration—was an improvement over the preceding one. But none of them could have become established if the inadequacies of the previous device had not brought the organisms to impending starvation. Evolution works by threats, not by promises, yet its result is the opening of greener and greener pastures."

Still another result is larger and more complex organisms made possible by the transmission of the same vital hereditary information from one cell to its descendants. But only by changing these genetic code-carrying molecules was it possible to produce enough change to create a variety of species. Even a slight alteration in the gene in either the arrangement or the number of its molecules will change the message carried by that gene. Thus, catastrophe, in the form of radiation, chemical insult or other mutagen, changes the genetic message and is an essential agent of evolution.

Dr. Immanuel Velikovsky sees catastrophe as not merely *an* agent of evolution, but as *the* agent. "In order for a simultaneous mutation of many characteristics to occur, with a new species as a resultant," he wrote in *Earth in Upheaval,* "a radiation shower of terrestrial or extraterrestrial origin must take place. Therefore, we are led to the belief that evolution is a process initiated in catastrophes. Numerous catastrophes or bursts of effective radiation must have taken place in the geological past in order to change so radically the living forms on earth, as the record of fossils embedded in lava and sediment bears witness."

The actual process of genetic evolution, wherein an original, ancestral gene is changed into something else, has now been studied in the laboratory and was recently described by Dr. Clement Markert, professor of biology at Yale University. "The evolution of a gene," he wrote in *Science* magazine, "involves two distinct processes. First a new gene arises by duplication from an original gene and then it diverges from that gene by the accumulation of mutations which alter its structure and correspondingly the structure and function of its product. The second part of the evolutionary process involves changes in regulation so that this new gene is expressed at those times and in those cells for which it is advantageous and is not expressed when it would be detrimental."

The biochemistry of genetic catastrophe is now reasonably well understood and has given rise to a new controversy, bringing into question once more the ability of Darwinian evolution to explain all the changes that can be observed in nature. For where once naturalists such as Darwin looked for signs of evolutionary change by counting feathers and measuring bones, population geneticists now search for amino acid substitutions and protein polymorphisms. In them, they profess to see not the grand design of nature, but a wholly random selection of new genetic forms, mutations created by a blind crapshooter throwing sevens and every other point for no apparent reason. As an example, the neutrality theory, as it is called, points to the different blood types of both human beings and chimpanzees. Blood is composed of a variety of proteins, and the characteristic types—A, B, O and so forth—are constructed from different amino acid sequences. There are twenty amino acids used by living things to build protein, and the sequence in which they are arranged determines the type of protein assembled. In the case of differing blood types, the presumption is that they are all mutations derived from a single, original amino acid sequence. The different mutated forms have since become fixed in the gene pool but offer no apparent advantages or disadvantages to the individual or to the species.

This, say the neutralists, is because mutations occur on the molecular level in a totally random fashion and are not affected by the principles of natural selection. As a major example, they point to the fact that several animal species—birds, bats, guinea pigs and humans—have lost the ability to synthesize vitamin C. Traditional Darwinian evolution argues that the loss must in some fashion be an adaptation to some special environmental pressure or circumstance.

The adaptive value of the lost capability is that animals with diets rich in vitamin C would save biochemical energy for other uses by not having to produce it themselves.

Not so, say the neutralists, and they point to many grazing animals with vitamin C-rich diets that retain the ability to make it themselves—a wasteful and disadvantageous trait according to natural selection. Comparing birds and mammals from various lines of evolutionary descent, the neutralists find that some have lost the trait, while others have retained it, and in neither case does there seem to be any evolutionary advantage or penalty.

Differences such as these between individuals and species are the product of many small mutations that involve nothing more complex than the substitution of one amino acid for another within the framework of a large protein molecule.

We can see these substitutions with the aid of a relatively new technique called electrophoresis. Proteins are put in an electrical field and will drift to either the negative or the positive side of the field, depending on their inherent charge. When two or more proteins are placed in the field, each will move different distances at different rates. This combination of rate and distance of movement is distinct for every protein, giving it its own fingerprint, or electrophoretic signature.

Electrophoresis thus provides a means of analyzing and identifying individual genes. Electrophoretic signatures have shown a great variety in the structure of proteins that carry out the same tasks in different individuals of the same species. This is known as protein polymorphism. Hemoglobin, the protein that carries oxygen in the blood, for example, in two different human beings has two different electrophoretic signatures.

"Every individual in any living species is different genetically from every other," explains Harvard University geneticist Professor R. C. Lewontin, who helped develop the technique of electrophoresis. "All people are genetically different from one another in many, many genes."

These polymorphic genes are the product of mutations. There is no question among biologists about that. The delicate coding structure can be easily and slightly altered by radiation or other mutagens in the environment or even by something known as error catastrophe, a copying error when the code is transcribed from blueprint to building of the protein.

What biologists have begun to question is whether or not natural selection works on the molecular level. In the electrophoretic signature they hoped to see either Darwinian design or, as the neutrality theory suggests, nature shooting craps.

"What we are really talking about on the molecular level," says Dr. Jack L. King of the University of California at Santa Barbara and one of the authors of the neutrality theory, "is whether all of the fine detail of a protein is controlled by natural selection, or whether you have gone beneath the resolving power of natural selection and are looking at random changes."

The question has not been answered by electrophoresis as yet, for it has come up with some rather puzzling results. Biochemical comparisons of forty-four proteins and enzymes common to both chimpanzees and human beings show them to be 99 percent identical. When DNAs of the two species were compared, the differences were considered too small to account for the substantial anatomical, physiological and behavioral differences that exist between human and chimp.

How then did mutation separate human from ape?

Studying the problem, biochemical geneticists Drs. Mary-Claire King and A. C. Wilson of the University of California, Berkeley, concluded "that the genes of the human and the chimpanzee are as similar as those of sibling species of other organisms." The significance of this comment is that sibling species are often so similar in appearance that it takes careful biochemical analysis to tell them apart.

Two species of frog that are similar in anatomy and behavior, for example, may have protein differences thirty to forty times greater than those between human being and chimpanzee. How is that possible?

"We suggest," Drs. King and Wilson offer, "that evolutionary changes in anatomy and way of life are more often based on changes in the mechanisms controlling the expression of genes than on sequence changes in proteins. We therefore propose that regulatory mutations account for the major biological differences between humans and chimpanzees."

In other words, wholesale changes in the genetic structure were not needed to produce the appropriate mutations that led human and ape on dissimilar evolutionary paths some 20 or 30 million years ago. Rather, it might have been a scrambling or rearrangement of the genes on the chromosome that placed one set of genes next to a different regulator gene. This would alter the timing and duration of their activity, but the genes themselves need not have been changed.

Such changes in the timing of the gene's activity might cause bones to grow longer or change their shape. It could ensure that fur continues to grow or stops altogether or that brains continue to grow until they reach the size of human brains or do not. The genetic message has not changed; only its expression has. Bone remains bone, hair remains hair, and brain cells are still brain cells. Only their size or number may change, and it is these differences that are responsible for species differences.

According to this theory, profound evolutionary changes were produced by seemingly minor catastrophes within the cell, catastrophes that may not have affected the structure of individual genes but simply rearranged their order within the chromosome. From such minicatastrophes were born the profound changes that sent the human on a very different evolutionary path from his ancestral species cousin the ape.

Now another catastrophe looms to set evolution on still another course. This one, however, will not submit to a time-consuming process of trial and error to be picked over by natural selection. Rather, it will be accomplished with stunning swiftness and not by the forces of nature, but by the hand of man. For we have now reached the point where we are no longer content to observe and understand our own evolution; we are now attempting to direct it.

"The age of synthesis is in its infancy, but is clearly discernible," wrote biologist James F. Danielli, of the State University, New York, at Buffalo, in the *Bulletin of the Atomic Scientists,* the thinking person's version of a doomsday book. Danielli was commenting on what is popularly known as genetic engineering, the construction and alteration of genes in the test tube.

"Previously, plant and animal breeders have been able to create what are virtually new species, and have been able to do so at a rate which is of the order of $10^4$ [10,000] times that of average evolutionary processes," Danielli points out.

But with the development of a few new techniques, Danielli foresaw a rapid increase in the rate of human-caused evolutionary change. "When these techniques are available, the possible rate of formation of new species will again be accelerated by a factor of $10^4$ to $10^5$. All existing genes, and all genes which do not presently exist, but which can be synthesized, will be available for the synthesis of new organisms. It will be possible to carry out the equivalent of $10^8$ to $10^9$ [1 billion] years of evolution in one year. It will be surprising if we do not reach this point within 20 to 30 years, and we may well be there in 10 years."

Danielli wrote these predictions in December 1972. In December 1975 a team of Harvard University biologists created an entire mammalian gene.

This followed, by only five years, the synthesis of a very simple yeast gene. It was simple in that the DNA of the gene was composed of only seventy-seven code letters, or nucleotides. In the latest gene synthesis the Harvard researchers created the gene for rabbit hemoglobin—a complex molecule containing 650 nucleic acids strung together. It is eight times larger than the yeast gene and as long as a human gene.

The Harvard technique uses messenger RNA to copy the gene's DNA. This is simply a reversal of the normal genetic process in which DNA orders the production of a messenger molecule known as mRNA to transmit the proper orders to the amino acid assembly machinery of the cell. An enzyme known as reverse transcriptionase reverses that process and makes the synthesis of complex animal genes possible.

It also means that almost any gene can be built in the test tube. "You can use this approach to make a double-stranded DNA molecule from any mRNA molecule you can purify. The number of genes you can isolate is simply proportional to the amount of mRNA starting material," explained Thomas Maniatis, a member of the Harvard team.

The ability to manufacture, transplant and combine genes is at the heart of a potential catastrophe that threatens the entire human race with a biological doomsday. Biologists, for example, can now cross genes of different species to create live hybrid cells. Researchers at the Swedish Institute for Medical Cell Research and Genetics recently combined human cells with chromosomes of other animals such as rats, mice and even insects and unborn chicks. The end products are "man-mouse" cells and other even stranger combinations. Equally strange was the new hybrid cell's biological behavior, which was often totally unpredictable.

Another technique in use at hundreds of laboratories all over the world allows genes from any source—viruses, bacteria and animal cells—to be implanted into common and ordinarily harmless bacteria known as *Escherichia coli*. *E. coli,* as they are more familiarly known, are normally found in the human digestive tract, where they aid in the digestive process. In the laboratory the *E. coli* have become the workhorse of genetic transplant experiments because of their size and tremendous ability to reproduce rapidly. The *E. coli* also contain DNA molecules, not only within their nucleus, but also in a readily accessible form, floating free in the cytoplasm of the cell.

Called plasmids, these strands of DNA can be lifted from the cell and transplanted into a new cell, where they are perfectly at home and divide and reproduce each time the cell divides and reproduces. Plasmids become far more than just a laboratory curiosity, however, for they can be neatly sliced apart by a biochemical known as a restriction enzyme. Other DNA molecules can then be stuck onto the cut ends of the plasmids, creating a new and very different gene. Then these so-called recombinant genes are stuffed into *E. coli* and give them wholly new and sometimes very dangerous biological characteristics.

"Several groups of scientists are already using this technology to create recombinant DNA molecules from a variety of other viral, animal, and bacterial sources," says Dr. Paul Berg, chairman of Stanford University's Department of Biochemistry. "These experiments could result in the creation of novel types of infectious DNA elements whose biological properties cannot be predicted. The new infectious DNA elements could spread widely and quickly among humans with unpredictable effects if they escaped from the laboratory."

The specific fear is that self-reproducing bacteria such as *E. coli* which are infectious to man might be so altered genetically that they would resist all medical treatment. The specter of worldwide epidemics similar to the 1917–1919 influenza pandemic, which killed an estimated 30 million people, and of tumor-causing viruses raging like wildfire throughout the world looms large in the nightmares of some scientists working in the field.

What makes these new infectious agents even more frightening is that they may not produce any immediate symptoms, and by the time they do it will be too late. In the past researchers working with infectious microorganisms who became ill as a result always developed known symptoms. "But now," warns Nobel Prize-winning biologist Dr. David Baltimore, "we may be talking about infections that might occur without immediate effects and therefore could spread among the population without the signal of illness. This presents a different kind of danger."

Laboratory workers infected by these new and potentially deadly genes would go home from work, infect their families and friends and trigger a monstrous epidemic against which there would be no controls. For in most cases these new bacteria are resistant to all known antibiotics. Some contain not one but a multiple capacity for harm. Some genes may carry programs for the creation of poisons that will destroy the host, or they may interfere with the regulatory mechanism of the cell, preventing it from dividing and permitting it merely to grow and grow—in effect turning it into a cancer cell.

Nor is the creation of a swarm of "Typhoid Marys" the only means of putting deadly new strains of infectious bacteria into the environment. Standard practice at many laboratories is to pour used and even viable cultures in which the dangerous bacteria live down the laboratory sink. From here the bacteria soon find their way into the municipal sewer system and thence into the food and water supply.

The hazards posed by these recombinant genes became increasingly manifest to Paul Berg when his own experiments took him onto what he believed to be highly dangerous ground. In 1972 he managed to create a hybrid DNA molecule which contained a malignant virus that caused tumors to develop in monkeys. Not interested in its cancer-causing potential, Berg expected the virus to produce interesting new properties when introduced into an *E. coli* bacterium. At that point he faced the chilling question of what would happen if the *E. coli* carrying a monkey tumor virus should escape and infect the population outside the laboratory.

"I decided not to do the experiment," recalled Berg, "because I couldn't persuade myself that there was zero risk."

At the time Berg's technique for creating the hybrid molecule was fairly sophisticated, but then other scientists at Stanford and the University of California at San Francisco developed a more simplified technique to introduce frog genes into DNA.

"That upped the ante," said Berg, "because it showed how simple it was to introduce any gene you liked into bacteria."

The possibilities for biocatastrophe are now vastly multiplied because the technology needed to turn harmless *E. coli* into billions of microbial Frankenstein monsters is so simple it can be done by almost anyone.

"Restriction enzymes," says Dr. Berg, "made gene recombination simple enough to be done in a high-school science class or on a kitchen table. When it became clear what could be done, we began to get calls from people asking for the enzymes so that they could do all sorts of things."

So concerned was Berg that he and a number of colleagues called for a special conference to draw up guidelines to safeguard the human race against the frightening possibilities of the new genetic engineering.

In July 1974 the first meeting of the Committee on Recombinant DNA Molecules took place and called for an embargo on two types of experiments. Type I experiments involve the insertion of bacterial genes which confer either resistance to antibiotics or the ability to form bacterial toxins. Type II experiments involve the insertion of viral genes into bacteria.

A third type of experiment—inserting animal genes into bacteria—"should not be undertaken lightly."

The suggested embargo caused an uproar among molecular biologists, who felt that the research was of such great potential benefit that it should not be halted, and among others who saw it as a threat to scientific freedom.

Nonethless, the potential hazard was viewed to be so great that the Berg group called for an international meeting to establish guidelines to replace their proposed moratorium.

In February 1975, 139 biologists from seventeen countries met at Asilomar, California. They considered the various types of experiments, their potential risks, benefits and the possible means of control. The implementation of their general considerations was left to the appropriate agencies in the various nations. In the United States it fell to the National Institutes of Health, which established an advisory committee on recombinant DNA research. The committee designed a series of guidelines to safeguard the public from what could be a horrendous biocatastrophe. The safeguards are primarily aimed at containment of the experimental material by both physical and biological means. Where the experiment is considered absolutely safe, no special techniques beyond normal laboratory precautions are considered necessary. For more dangerous experiments laboratories will be equipped with air locks and will be contained by negative air pressure, and all waste materials will be decontaminated.

The assignment of risk was made on the basis of the committee's assessment of the dangers to humans and nature that could result from "the worst possible scenario" for each type of experiment. Those involving DNA taken from primates must have high containment. As the experimental material drops down the phylogenetic scale from mammals through birds, cold-blooded vertebrates, invertebrates, etc., the containment levels also decrease. Higher plants require more stringent controls than lower orders. When pathogenic hosts, or vectors, as the material carrying the DNA into the cell is called, are used, containment levels are raised.

In addition, emphasis has been placed on the development of "safe bugs," microbial organisms that cannot survive outside the laboratory.

The task these biologists have set for themselves is enormous, and never before in history have scientists voluntarily attempted to police themselves and back away from what they consider their inalienable right to seek truth, wherever it may take them.

But there are those obviously who recognize the danger and fully appreciate the social responsibility the scientist must assume. Scientists, according to Dr. Robert Sinsheimer of the California Institute of Technology, "must finally recognize that to reshape man is not a beguiling laboratory experiment, but an enterprise that involves the ultimate in value judgment."

But who can withstand the lure of potential benefits scientists see once they have taken over from nature the task of directing evolution? Once the technology is fully in hand, researchers plan to modify normal organisms and correct defective ones. Plant physiologists see the time when staple food crops can be outfitted with all the most efficient plant equipment: genes to fix nitrogen, resist disease, produce essential amino acids and increase carbohydrate production. Medical researchers believe they will be able to correct genetic errors that cause human and animal diseases by cutting out defective genes and replacing them with normal ones.

But the specter of biocatastrophe hangs over these incredible benefits. Never before have the weapons for germ warfare been easier to create and more difficult to defend against or control. Despite all the guidelines, there is still no real means of preventing research that could lead to the most dangerous hybrid creations and no means of preventing them from escaping inadvertently or willfully into a helpless world.

There is a bizarre irony in all this. The biological revolution that seeks to take control of evolution was begun as an attempt to replace catastrophism with a less mythical understanding of natural processes. Our understanding has increased far beyond the wildest imaginings of the uniformitarians, and with it our ability to control nature. The irony is that by ruling out catastrophe as a natural, creative force, we may unleash a biocatastrophe far worse than any disaster nature could achieve.

# 8

# INTO THE SOUP—CREATION BY CATASTROPHE

It is curious, but nonetheless true, that scientists are excellent observers, but on occasion they seem even more accomplished at denying the evidence of their observations. Nowhere is this more true than where the phenomenology of life's origin is concerned. Although the creation of life is permitted to be godlike, it must also be both plausible and mechanical. Never, however, is it permitted to be poetic. Thus, the concept behind Aristotle's observation that fireflies arise out of the morning dew was one of the urgent targets of nineteenth-century science. Not for its fey qualities, but rather because it suggested that life could arise from nonlife. This concept known more formally and less poetically as spontaneous generation persisted until the middle of the nineteenth century.

The formulas for the creation of certain forms of life were well known. Mud, silt, sweat, excrement and the decaying parts of dead organisms were among the most common ingredients. Shakespeare in *Antony and Cleopatra* names them as stuff of life.

> Your serpent of Egypt is bred now of your mud
> By the operation of your sun: so is your crocodile.

Modern biology has, in fact, not moved terribly far from that Shakespearian view of the origin of life. Famed biologist Salvador E. Luria writes glowingly of the role played by clay in the creation of life: "Almost certainly, clays played an important role: chains of amino acids can be produced in the test tube in the presence of certain types of clay. When seen through the prism of science, the biblical reference to the moulding of man out of clay, 'I also am formed out of the clay,' Job 33:6, may contain an element of truth. Clays are remarkable substances, made up of microscopically thin plates with water-seeking chemical groups on each side. These chemical groups are probably the catalytic agents—the pre-enzymes of the distant past—relatively unspecific yet of tremendous use in the creation of life."

A far more specific recipe for the creation of life is contained in a seventeenth-century Belgian textbook. It noted that body lice, bugs, fleas and worms are born "from our entrails and excrements." It also suggested a method for the spontaneous generation of mice by putting a piece of sweat-soiled underwear together with some wheat in an open jar. After three weeks, it advised, mice of both sexes will be found in the jar. The actual mice-making process is accomplished by the "ferment" of the sweat emerging from the soiled underwear to penetrate the husks of the wheat and thus change the wheat into mice.

For a time the church went along with the idea of abiogenesis, at least where lower forms of life were concerned. St. Augustine declared that spontaneous generation was a means by which God demonstrated His omnipotence, by interfering with the usual orderly sequence of events. But the Renaissance, which shook church dogma at a number of points, also weakened its faith in the spontaneous generation of life.

For while one authority suggested stuffing underwear into a flask to create mice, another expert, a Tuscan named Francesco Redi, was busy demonstrating that the maggots that suddenly wriggled into existence on rotting meat were nothing more than fly larvae. As court physician to Ferdinand de' Medici, Redi was in so strong a position he could challenge church doctrine and demonstrate his idea that it was the flies that had deposited their larvae in the meat and not the Lord.

Adding to the confusion was a discovery by a Dutch lens

grinder named Anton van Leeuwenhoek, who saw in the newly discovered microscope "animalcules," creatures as alive and active as a horse, yet small enough to live in a droplet of lake water. Van Leeuwenhoek was probably the first man in history to see the one-celled creatures we know as protozoa.

Such tiny creatures, he believed, could not have been born in the usual sense and must certainly have simply popped spontaneously into existence.

And so the argument raged. The church and most of the scientific hobbyists of the day were prepared to accept Redi's evidence, and agreement was reached that all animals visible to the naked eye could not arise spontaneously. They required parents, and even insects were believed to hatch from eggs and not simply pop into existence. But the microscopic "animalcules" that Van Leeuwenhoek had found at the other end of his lens were another matter.

The church, having been caught out in scientific left field so many times before, was anxious to hedge its theological bets and also willing to accept the evidence of experimentation that would resolve the question one way or the other. A number of priests began investigating the question. One was John Needham, an English naturalist who was also an ordained Roman Catholic priest. In 1748 he boiled up a batch of mutton broth and sealed it in a glass jar. A few days later he opened the jar and looked at the broth under a microscope. It was teeming with microorganisms, which Needham called infusoria.

Now the church had hard evidence of spontaneous generation, and a neat theological line was drawn. All those creatures that people could see with their own unaided eyes simply could not be the product of anything except other living things. Those "infusoria" and other "animalcules" that they could see only with the aid of the microscope were created out of inanimate objects and did not require living parents to come into being.

But there were those who simply could not let well enough alone. Twenty years after Needham's experiments had allowed the church to set its theological house in order on the question of abiogenesis, along came another priest, one Lazzaro Spallanzani, an Italian, who repeated Needham's experiments with very different results. He sealed broth up in jars and boiled them for a very long time, up to one hour. Then the jars were left standing for several days. When they were examined under the microscope, the boiled broths contained no "infusoria," no "animalcules"; in short, they were empty of life.

Spallanzani's results, however, were suspect on two counts. The first was the fact that he had become a priest not out of any great religious calling, but rather to help support himself since at that time most university teaching posts in the sciences went to priests. No one was ungracious enough to raise the point, however, and his work was questioned by Needham on the ground that Spallanzani had brutalized the air in the flasks and thus had destroyed its ability to sustain life.

There matters stood until 1864, when the French Academy of Sciences offered a prize for a convincing experiment that would answer the question of spontaneous generation one way or the other. The apparent winner was Félix A. Pouchet, director of the Museum of Natural History in Rouen, France, who had carried sealed flasks filled with decaying hay to the tops of the Pyrenees and there opened them. The contents of each flask had begun to ferment when opened, proof, according to Pouchet, that the bacteria responsible for the fermentation had been spontaneously generated.

Not so, said Louis Pasteur, who had previously performed similar experiments and got no fermentation. Pasteur accused Pouchet of contaminating his flasks by breaking them open with nonsterile nippers.

To settle the controversy, the French Academy proposed a joint experiment to be done by Pasteur and Pouchet. Pouchet refused. But on April 7, 1864, Louis Pasteur mounted the rostrum at the Sorbonne to deliver the results of yet another of his experiments designed to prove that spontaneous generation was impossible. The controversy was generated not only by the subject but also as a result of Darwin's *Origin of Species,* which had been published only five years before. The church had taken its position against evolution, and Pasteur, a devout Catholic, felt duty bound to support it. That April night in Paris, Pasteur told his listeners at the Sorbonne that he would direct his remarks to one aspect of the evolutionary debate—namely, "Can matter organize itself on its own? In other words, can beings come into the world without parents, without ancestors?"

The controversy sparked by Darwin represented, according to Pasteur, "two great currents of thought as old as the world, and which in our time, are known as materialism and spiritualism. What a victory, Gentlemen, for materialism if it could be shown that matter can organize itself and come to life," said Pasteur with great emotion. "Ah, if we could give [to matter] that other force which is called life . . . what need to resort to the idea of a primordial creation, before whose mystery one must indeed bow down? What need for the idea of a God-creator?"

To Pasteur, of course, the need was obvious, and he was prepared to demonstrate it by proving once and for all that spontaneous generation was impossible. The spark that created the bacteria was bacteria themselves. Life could arise only from other life. Then he lifted a flask containing a fermentable material. The normally open neck of the flask had been twisted into an S shape and drawn out to a point with only a narrow tip open to the air. So narrow was the opening that little dust-containing microbes could enter, while the twisted neck would prevent any dust motes from falling into the material at the bottom of the flask. The flask had lain untouched for four years, and the material inside was unfermented—proof, Pasteur insisted, that spontaneous generation was impossible.

"No," he cried, "there is no circumstance known today whereby one can affirm that microscopic beings have come into the world without germs, without parents resembling themselves. Those who claim it are the playthings of illusions, of badly done experiments tainted with errors that they did not know how to recognize or that they did not know how to avoid."

Pasteur's experiment and demonstration at the Sorbonne effectively silenced all those who believed in spontaneous generation, although it did little to retard the spread of the idea of evolution. But even Darwin, who believed that life somehow had to arise spontaneously at some time, admitted that he knew of no experiment that could reliably demonstrate the event. He remained convinced that life itself must have evolved from nonlife, but determining the conditions and chemicals involved was beyond him. In yet another burst of intuitive genius, Darwin realized that the chemicals essential to the life process cannot exist in a world dominated by living things. Outside the living system such chemicals are instantly consumed by the living organisms that utilize them. But, thought Darwin, suppose the chemistry of the earth had been different in the past from today?

"It is often said," he wrote, "that all the conditions for the first production of a living organism are now present, which could ever have been present." This is pure uniformitarianism, but Darwin was prepared to make certain exceptions, certain departures from the rigidity of Lyell's rules. "But if (and oh! what a big if!) we could conceive in some warm little pond, with all sorts of ammonia and phosphoric acid salts, light, heat, electricity, etc., present, that a protein compound was chemically formed ready to undergo still more complex changes, at the present day such matter would be instantly devoured or absorbed, which would not have been the case before living creatures were formed."

With this statement Darwin was reinforcing the essential unity of all life which is the cornerstone of modern biology. But Darwin's Victorian metaphor of tranquility, wherein a handful of chemicals came gently together to form life, is far removed from what actually happened. Life is the product of a series of violent events that can be traced back to the catastrophic death of the stars. In the course of a star's life it will consume virtually all the hydrogen of which it is composed. It does so by a process known as fusion, in which the protons of the hydrogen atom are fused together to form a helium atom. This nuclear burning continues for 99 percent of the star's lifetime. Then, with the hydrogen virtually consumed, the star's core begins to collapse until the temperature at the center reaches about 200 million degrees F. At this incredible temperature the helium nuclei begin to fuse, forming carbon nuclei. The collapse then continues, and oxygen and still heavier elements are formed, until all the elements of the universe have been created from the original hydrogen building block.

If the star's mass is small, such as that of our sun, its outer shell continues to fall in upon itself, until its entire mass has been squeezed into a volume the size of the earth. With a density of 10 tons per inch, such highly compressed stars, known as white dwarfs, simply radiate away what little remains of their heat and fade into darkness.

"A different fate awaits a large star," explains Dr. Robert Jastrow, director of NASA's Goddard Institute for Space Studies. "Its final collapse is a catastrophic event which generates temperatures of several billion degrees, burning the last residue of fuel sprinkled throughout the star, and releasing a burst of energy which blows the star apart. The exploding star is called a supernova.

"The supernova explosion sprays the material of the star out into space, where it mingles with fresh hydrogen to form a mixture containing all 92 elements. Later in the history of the galaxy, other stars are formed out of the clouds of hydrogen which have been enriched by the products of these explosions. The sun, the earth, and the beings on its surface—all were formed out of such clouds containing the debris of supernova explosions dating back billions of years to the beginning of the Galaxy."

Recently astronomers have begun to find that some of the compounds formed in the aftermath of supernova violence were, in fact, the raw materials from which life is formed on earth. Formaldehyde, ammonia, hydrocyanic acid and other molecules known to nineteenth-century chemists as "organic" compounds, because they could be created only by living things, fill many regions of space. When the compounds collide, they send radio signals through space; each molecule has its own distinctive wavelength. Thus, molecular astronomers have been able to identify many organic chemicals floating about among the gas and dust of the galaxy's spiral arms. In the ongoing process of star birth and death, these components, as well as the hydrogen and other elements packaged in the dusty arms, are incorporated into the planets and stars of new solar systems.

In just such a fashion our solar system originated, and the materials from which life might be created were similarly incorporated into the primordial earth.

Spinning about the sun some 4.5 billion years ago, rocked by volcanic eruptions and earthquakes, the earth began to generate an atmosphere of fiery gases squeezed from its solid mantle by great currents of heat produced by radioactive decay. That process still continues inside the earth's core.

"The atmosphere, at that period," wrote the Russian biologist Alexander Oparin in 1924, "differed materially from our present atmosphere in that it contained neither oxygen nor nitrogen gas, but was filled instead with superheated aqueous vapor. . . . The superheated aqueous vapor of the atmosphere coming in contact with the carbides reacted chemically, giving rise to the simplest organic matter, the hydrocarbons which in turn gave rise to a great variety of derivatives."

What Oparin postulated was the revolutionary idea that had already been effectively disproved by Pasteur and others seventy-five years before: that life could arise from nonlife, that individual inanimate chemical molecules could combine and evolve into a living organism. To Alexander Oparin then must go the credit for developing, as one recent biochemistry textbook declares, a "convincing formulation of the concept of *prebiological organo-chemical evolution.*"

The details, of course, have been modified somewhat, but the Oparin theory has been basically substantiated.

The idea of a primitive atmosphere giving rise to life was further refined by the British biologist J. B. S. Haldane, who saw in 1929 the mechanism by which life could be created on earth. Haldane had made a great but logical leap in concluding that if there were no oxygen in the earth's early atmosphere, there would also be no ozone shield, and the earth would thus be bombarded by ultraviolet light. "Now," wrote Haldane in *The Rationalist Annual,* "when ultraviolet light acts on a mixture of water, carbon dioxide and ammonia, a vast variety of organic substances are made, including sugars and apparently some of the materials from which proteins are built up. . . . In this present world, such substances, if left about, decay—that is to say, they are destroyed by microorganisms. But before the origin of life they must have accumulated till the primitive oceans reached the consistency of hot, dilute soup."

Could it actually have happened that way? Was life formed from a bunch of gases stirred by ultraviolet energies into a rich broth from which ultimately emerged a creature capable of questioning its origins? Biologists are not quite in agreement, especially with the hot soup idea. That is basically a closed system, and life is presumed by most biologists today to be a collection of open systems which derived from an open system. But all the speculation of Oparin, Haldane and others, despite the occasional wrong turnings, cried out for experimental confirmation, and that was finally forthcoming in 1953, when Nobel Prize-winning chemist Dr. Harold Urey put a graduate student to work on the problem at the University of Chicago. His name was Stanley Miller, and he constructed a sealed system of flasks and tubes into which he injected methane, ammonia, hydrogen and water vapor—the gases Urey believed constituted the earth's primal atmosphere.

This chemist's brew was then stabbed with 60,000 volts of electricity—a sort of man-made lightning.

After a week of this laboratory gestation the collection flask at the end of the system was filled with water that had turned deep red. When analyzed, it was found to contain several organic acids, including amino acids. There are eighty known amino acids; but only twenty are found in natural protein molecules, and they are the only ones considered the building blocks of life.

The Urey-Miller experiments produced four of those twenty amino acids and launched a new field of investigation: the creation of life. Other experimenters varied the gas mixture slightly and produced other amino acids. Soon various experimenters were producing eight and ten amino acids, but no one successfully managed to produce all or even most of the twenty amino acids in one single experimental genesis.

A great deal of controversy surrounded these atmospheric experiments, with each group certain that the gases it proposed were, in fact, the actual ones present at the time. All, however, included free hydrogen, the first and basic element of the universe, in the primal mix. Then, in 1964, at Florida State University, Dr. Sidney Fox proposed a primal atmosphere that had little or no free hydrogen in it, to account for the amino acids that were very poor in hydrogen—precisely those, in fact, that had never been produced in all the other atmosphere experiments.

Fox and his co-worker, Dr. Karou Harada, put together an atmosphere of methane, ammonia and water, with no free hydrogen to react with the intermediate products being formed from the gases on their way toward becoming amino acids. They also used a different kind of energy input from those of earlier experiments. Miller had used electrical discharges; others had bombarded the gas with alpha particles and ultraviolet radiation. Fox used heat. It was, he believed, the energy source most readily produced by the geological processes then going on in the early earth. "Volcanic activity and other conditions," he points out, "indicate that the earth was much hotter than it is now, with thousands of hot zones well above the boiling point of water."

The result of the experiment was amino acids in profusion; twelve amino acids were formed, and each was an amino acid common to protein production.

Although the controversy over the primitive atmosphere continues, all the debaters are agreed that one element was certainly not present then—oxygen. This lack was as important to the creation of life as any phenomenon that did take place, for if oxygen had been present, the fledgling molecules would have simply ended up as the waste products of combustion, and not as the precursors of protein.

The lack of oxygen also meant that ozone, a heavy form of the oxygen, was also lacking and that the screen it now provides against most of the fierce ultraviolet rays of the sun was not present. The primordial atmosphere was bombarded every moment of daylight by ultraviolet radiation, a condition that would be lethal to living organisms, but that served as an energy source for the creation of life.

"Most of this chemistry," explains Dr. George Wald of Harvard, "probably took place in the upper reaches of the atmosphere, activated mainly by ultraviolet radiation from the sun and by electric discharges. Leached out of the atmosphere over long ages into the waters of the earth, organic molecules accumulated in the seas, and there interacted with one another, so that the seas gradually acquired an increasing concentration and variety of such molecules."

Here again we see the view of the uniformitarians, the idea of slow, gradual development and evolution. To these gradualists, all that was required at this point was the slow, stately procession of time to provide the opportunity for small molecules to link up to form the far more complex amino acids that are in turn the building stones of protein. But some scientists argue that even nature must bow to some chronological pressures to speed things up.

Could such a hurry-up process have been catastrophic

in nature? Could a series of violent events have rocked the earth, roiled the atmosphere or boiled the hot dilute soup that Wald and Haldane say were its oceans to the point where the molecular precursors of life exploded into life?

One mechanism for the genesis trigger might have been shock waves produced by giant thunderclaps or even by meteors plunging into the atmosphere. At Cornell University a pair of Israeli researchers, a husband-and-wife team, Akiba and Nurit Bar-Nun, tested the theory. They filled one end of a tube with a mix of the earth's presumed primal atmosphere—ammonia, methane, ethane and water vapor. The gases were separated by a slim plastic membrane from a layer of inert helium at the other end of the tube. The pressure on the helium was increased until the membrane burst. This generated a shock wave that ripped at high speed through the gas mix in the other end of the tube. The fast-moving shock wave raised the temperature inside the flask by several thousand degrees, if only for a brief moment. When the shock wave passed and the temperatures had dropped back down to their normal levels, four amino acids had been created.

Compared with the experiments that used ultraviolet radiation to heat the brew, the shock waves produced 36 percent more amino acids. The reason: The temperature rises produced by shock waves were too short-lived to break up any of the newly formed amino acid molecules.

Other experiments also point to catastrophic heating as the mechanism by which the chemical precursors of life were generated.

Consider the following scenario. We know the original atmosphere of the earth was different from that of today. There is also considerable evidence to indicate that early in the earth's history the solar system had not yet settled down into its present form. Velikovsky, of course, insists that the present form is of very recent origin, but our concern in this instance is with the earth-moon system. Today the moon is about 240,000 miles from the earth—three days by Apollo spacecraft. But there is some indication that about 3 billion years ago the moon was much closer to the earth.

Some evidence is contained in the shells of certain

coral and primitive algae. Their growth patterns are determined by the length of the solar day. Studies of their fossils show that 3 billion years ago the length of a day on earth was considerably shorter than it is now. So much so that 420 of those days constituted a solar year. This meant that the earth was rotating much faster then than it does now. According to the laws of angular momentum, which govern the earth-moon system, that faster rotation of the earth calls for the moon to have been much closer. Thus, astronomers have calculated that about 3 billion years ago the moon was only about 80,000 miles away from the earth.

This proximity of the moon has led another group of Cornell University scientists, D. L. Tuncotte, J. C. Nordmann and J. L. Cisne of the Department of Geological Sciences, to speculate on its contribution to the creation of life on earth. With the moon that close to the earth, they postulate the gravitational pull would have sent enormous tides coursing through the oceans. Even today the combination of lunar gravity and earth's rotation produces tides that in areas like Nova Scotia's Bay of Fundy rise and fall 40 and 50 feet twice a day. Three billion years ago the tides were even more fantastic, and moon rise and set were cataclysmic occurrences. Twice each day the oceans were pulled virtually out of their beds and then dropped back, sending enormous waves of energy surging over the earth. That energy was quickly converted to heat, enough heat to vaporize the oceans.

The Cornell group points, in fact, to a surge of high-temperature vulcanism that took place about 2.8 billion years ago. Many rocks bear the marks of such high-temperature intrusions, the invasion of older rock formations by molten rock, owing to a sudden heating. Such intrusions, the scientists feel, could have been the result of tidal heating.

The "implied catastrophe," they note, coincides roughly with the first records of life. These first records were the fossils of a group of spheres called *Eobacterium isolatum,* dated at 3.1 billion years old. These primitive life forms may have been created in scalding waters, "which," say the Cornell scientists, "is consistent with the idea of thermal catastrophe. It seems within the realm of possibility," they conclude, "that a global thermal event might have been involved in the origin of life."

Was it heat then, the product of catastrophic tidal waves crashing back and forth over the earth and still more heat generated by the violent explosions of volcanoes as the superheated gases deep within the earth tore through the surface, that triggered life? And how long was the catastrophic fury required before life came into being?

Experiments performed in the laboratories of the Institute of Molecular Evolution at the University of Miami indicate that the sequence of events that led to life might not have taken very long at all. "Our experiments and the geological relevance of the conditions employed for them indicate that complexity arose from simplicity very rapidly and very often on the Earth," says Dr. Sidney Fox, who heads the institute. "The experiments, in fact, demonstrate that the sequence leading to a protocell would have occurred in a few hours."

The process itself might have followed the same mundane steps used to create such consumer goodies as nylon and plastic packaging materials. This, at least, is the idea of Dr. Fox. His method for the creation of life also bypasses the necessity for genetic instruction early in the game. He proposes that nature used a chemical reaction known as thermal polymerization as the process that probably linked amino acids together to form proteins. Industrial chemists have used polymerization to assemble related molecules into long chains called polymers, and this, in fact, is the basis of the plastics industry. In a similar manner, the living organism links up amino acids, first into a small polymer called a peptide, which consists of several amino acids hooked together. The more amino acids available, the more readily polymerization occurs. The peptides are then joined to form the extremely long chain molecules that are proteins.

This process of polymerization might be compared to the action of a zipper pulling together the sides of a jacket. It was this zipper chemistry that Fox proposed to use in the construction of a complex proteinlike compound he calls a proteinoid. "The conditions which existed to produce proteinoids are something we don't have to argue over, for they are here now," he says.

Fox reasoned that the chemical process of polymerization was an inevitable one. The tug which pulled both halves of a zipper together would, in polymerization, be provided by heat, and just as surely as the zipper came together when pulled, the amino acids would hook up to form proteinoids.

In putting the theory to the experimental test, specifically by heating amino acids, Fox violated a number of traditional chemical shibboleths. One was that if amino acids were heated above the boiling point of water, they would simply "go up in smoke." To get around this problem, he did not do what every biochemist does—put the amino acids in water. Instead, they were heated in their dry state to 150 degrees Centigrade (50 degrees above the boiling point of water), and the result was a "lightly colored material."

Chemical analysis showed it resembled protein molecules. Then came "God time." Fox bathed this new material, his proteinoids, in warm water. They took on a startling form, spherical shapes that could be clearly seen under the microscope. Moreover, there were hundreds of them in the field, bumping into each other in random motion produced by Brownian movement.

Fox, in what must rank among the most modest gestures in the history of science, called these objects microspheres. But they looked and behaved remarkably like living cells. While Fox was not prepared to call them alive, he did admit that they demonstrated a number of properties that were similar to those displayed by what he called contemporary cells.

"No other experiment," he declared, "has produced anything that even begins to approach the microspheres in their similarity to cells and the properties the cells have."

That catalogue of similarities shows the microspheres to be made of the same general structural material as true cells. They are uniform in size and fall within the same size range and shape as coccoid bacteria, which some scientists believe are the most primitive bacteria. Microspheres can be treated like living cells; they can be centrifuged or sectioned for examination. They can also be stained, a classical diagnostic test for bacteria. The microspheres also have a double-layered boundary similar to the walls of living cells.

So closely do the microspheres resemble living cells, in their outward forms at least, that some of the world's best biologists have been fooled. One distinguished scientist was shown pictures of two objects taken under an electron microscope. They were both oval-shaped with double-layered walls. "Which," the scientist was asked, "is the bacterium called *Bacillus cereus?*" The scientist chose the wrong one, declaring the microsphere to be the electron portrait of the *Bacillus cereus.*

The microspheres perform a number of what, to laymen, are startlingly lifelike feats. They do not simply lie placidly in the mother liquor from which they spring. They move about in their medium, and when one microsphere bumps into another, they communicate with each other! That is, they meet and establish a common tube that links one with the other; through this tube protoplasmic material flows from one microsphere to the other.

The medium in which they float is a total environment from which they extract nutrients and grow, much the same as an amoeba, a rosebush, a laboratory mouse and a human being.

The one feature of contemporary life forms that microspheres lack is genetic material, but they are still able to reproduce their own kind. Their method of reproduction is not, however, unique. Many cells divide in half to reproduce, a process known as binary fission. However, some yeasts and bacteria reproduce by budding, sending forth a balloonlike piece that swells and breaks off to become a new cell. Microspheres bud in a different fashion after a week or so in the liquor. The microsphere buds do not, however, break off by themselves. They must be dislodged by external means—by an electrical shock being sent through the mixture or by the solution being heated. The buds are then removed from the mother liquor and placed in a new solution.

When this solution cools, the proteinoid material is taken up by the buds and layered about its walls, much as an oyster deposits nacreous layers about a grain of sand. In this case the buds are the grains of sand layering the proteinoid about themselves. After about an hour of this the buds grow to the size of normal microspheres, and the process begins again.

Fox's work has precipitated great controversy and not coincidentally becomes the primal chicken or the egg question.

Which came first, the nucleic acids which subsequently formed into long chains of DNA and RNA to create genetic templates for cells to build upon or the cell in which nucleic acids were formed by mutation or other process and found to be of value and retained as the essential hereditary mechanism?

Could life in the form of primitive cells very similar to microspheres have originated on earth before the creation of genetic chemistry? Biologists are sharply divided over the issue, but Fox's experiments, which are conducted under the same geochemistry that existed on the earth when life was created, make a very strong case. "While it is easy to assume that molecular replication [of DNA] preceded cellular replication in evolution, the experiments show that the replication of a simple or minimal cell could have occurred with the utmost of simplicity, and therefore probably came first," says Dr. Fox.

While the question of which came first is undoubtedly important, the overriding significance is that it happened at all. Within the seething caldron that was the early earth, the stuff of the stars, liberally sprinkled about in the atmosphere and lithosphere, was churned, heated and thrown violently together to give rise to life. The earth has changed greatly since that time, and there is every expectation that still further changes are impending. Of greater import is the nature of those changes; will they be characterized by the slow, gradual change that is the hallmark of uniformitarianism, or will they be marked by violent, sudden catastrophe?

# 9

# THE DISORDERED EARTH

In the popular musical *Hair,* which relied heavily on astrology and nudity, the lyrics of the best known song aligned Jupiter with Mars. The alignment of Jupiter, not merely with Mars, but with all the other planets in the solar system, is not, however, mere fancy or astrological mumbo jumbo. Such an alignment of the planets, all dressed, eyes right, in a straight line on the same side of the sun, is a rare, but not an impossible, event. In fact, it last occurred over 150 years ago. What makes the so-called superalignment of the planets important, not only to astrologers who look for great events to be caused by or influenced by planetary alignments, but to everyone, is that some scientists are also afraid it may be the trigger for one of the greatest catastrophes ever suffered by the earth.

"In 1982," say physicists John R. Gribbin and Stephen Plagemann, who coauthored a theory published in book form as *The Jupiter Effect,* "the sun's activity will be at a peak; streams of charged particles will flow out past the planets, including the earth, and there will be a pronounced effect on the overall circulation and on the weather patterns.

"Finally, the last link in the chain, movements of large masses of the atmosphere will agitate regions of geologic instability into life. There will be many earthquakes, large and small, around susceptible regions of the globe. And one region, where one of the greatest fault systems lies today under a great strain, long overdue for a giant leap forward and just awaiting the necessary kick, is California. The situation is not directly comparable with that of 1809, the last time such a planetary alignment occurred, because we have no way of knowing how much strain the San Andreas fault was then under. Most likely, it will be the Los Angeles section of the fault to move this time. Possibly, it will be the San Francisco area, which has had a major quake. The prospect of both these sections of the fault moving at once hardly bears thinking about. In any case, a major earthquake will herald one of the greatest disasters of modern times."

Not everyone, needless to say, agrees with Gribbin and Plagemann. Their theory rests upon a vast increase in sunspot activity triggered, they believe, by the superalignment of the planets. Sunspots are relatively dark areas on the sun's surface—dark, that is, when compared with the brilliance of the rest of the sun, for sunspots still shine with a brightness equal to a hundred moons—which are thought to be created by turbulence deep within the sun. These spots occur in approximately eleven-year cycles and have very definite effects on the earth's weather, which in turn affects the rate at which the earth rotates.

Despite the awesome tables and computer analysis that Gribbin and Plagemann insist prove that sunspot activity increases with the superalignment of the planets, most solar physicists are not convinced their theory is correct. "There is nothing to the planetary theory of sunspots," says Dr. Don L. Anderson, director of the Seismological Laboratory at the California Institute of Technology. He does not, however, dismiss the idea out of hand, for it is not "on its face as ridiculous as it seems, but it does have serious weaknesses."

Chief among them is the belief that the planets influence sunspot activity. But scientists do not deny there is a connection between the triggering of earthquakes and changes in the earth's rotation rate. What is not clear is which is cause and which effect. *The Jupiter Effect* authors assume that changes in the spin rate of the earth are the cause and the earthquakes the effect.

About all anyone can agree on is that earthquakes are

awesome in their power. A major quake such as the one that ripped Managua, Nicaragua, in 1972, develops a force that is a thousand times more powerful than the bomb that fell on Hiroshima. More traditional measurements utilize the Richter scale, which defines the magnitude of an earthquake in terms of the motion produced in a standard instrument at a standard distance at a rate of one cycle every twenty seconds. On this scale, which takes its name from Dr. Charles F. Richter, the California Institute of Technology seismologist who devised it in 1931, any reading over 4.5 is considered potentially dangerous. The most severe earthquake ever to have occurred is thought to be the Chilean quake of 1960, which registered a magnitude of 8.6 on the Richter scale, or a release of energies in excess of 3 million Hiroshima-style A-bombs.

By plotting the waves of earthquake energies, scientists had come to the conclusion that Richter magnitudes level off at about 8.6 or 8.7. But two scientists at MIT say that such leveling-off effects may be deceptive. It is possible, they suggest, that earthquakes four times as strong as those recorded in recent years may be coming in the future.

"It is not clear," wrote MIT seismologists Dr. Michael Chinnery and Robert G. North, in *Science* magazine, "that the record of large earthquakes during the last 100 years is sufficiently detailed that the occurrence of such a catastrophic event can be ruled out."

Such superquakes may rip the earth as often as every fifty years, the two seismologists believe. Even earthquakes of lesser magnitude develop incredible forces that can, in a few seconds, reshape hundreds of square miles of the earth's surface.

In the Gallatin National Forest in southwestern Montana, the U.S. Forest Service has preserved, as a sort of geological morning after, 37,800 acres that were totally resculptured by an earthquake. The quake struck at midnight on August 17, 1959, and recorded 7.3 on the Richter scale. By contrast, the San Francisco quake of 1906 is believed to have been only somewhat stronger, perhaps achieving a magnitude of 8.2.

Before the quake, the Madison River Canyon attracted hundreds of fishermen and hunters each year. The river sparkled and shone as it rushed through the valley floor. Above towered a 7000-foot mountain, its slopes covered with pine and aspen. But beneath its tree cover the mountain was as spongy as a layer cake, composed of strata of soft, weathered rock held in place only by a ridge of hard dolomite.

The quake burst the top of the mountain as if it had been blown apart by an atom bomb. Eighty million tons of rock crashed down the canyon wall at the incredible speed of 100 miles an hour. The huge mass of trees, rocks and earth rocketed into the river, and the avalanche kept on going, until it finally came to a stop 400 feet up the opposite wall of the canyon. Choked with rock and earth, the river backed up, and within moments a lake began to form.

Great waves called seyches surged back and forth across the newly formed lake, alternately flooding one shore, then the other. Today, despite a spillway cut through the slide-formed dam, "Earthquake Lake" is 100 feet deep. Elsewhere huge fault scarps were sculptured along the face of the break, raising stony towers where there was once a tall mountain.

Today the raw edges of the land have been softened by wind and rain, while goldenrod and lupine carpet the valley floor. Deer and elk graze along the meadows that border the lake, and the Madison River that formed it still finds its way northward to join the Missouri. But the landscape is changed considerably from what it once was, and the catastrophic forces that remolded the Madison River Canyon are still there beneath the earth's surface and may be unloosed again at any time. In fact, earthquakes of varying intensity rock some portion of the earth every thirty seconds of every day of the year.

The earth is not a solid ball spinning majestically and deliberately about the sun along an undeviating orbit. Rather, we all are inhabitants of a spaceship racing around the sun at 66,000 miles an hour. In this passage the earth spins about an axis, a motion that produces day and night. But except at the equator, day and night are not divided equally, and each day the amount of daylight and darkness differs as the earth journeys around the sun. The forces that affect the earth's rotation around its axis are many and complex.

The earth's bulging equator sticks out somewhat like a fat man's paunch and offers a convenient hook for the gravitational fields of both the sun and the moon to grasp and then twist the earth as it moves through space. This shaking produces a distinct wobble in the earth as it spins about its axis. The poles shift as a result and do not always point to the same place in space. Changes deep in the earth itself, acting in concert with the global movement of vast masses of air, will also cause a wobble in the spinning planetary top.

The wobbles were first discovered in 1891 by an amateur astronomer named S. C. Chandler. One has a period of 428 days and is now named the Chandler Wobble in honor of its discoverer. The other wobble has a period of one year and is not surprisingly dubbed the Annual Wobble.

So what, so Spaceship Earth shakes a bit as it moves about its axis, surely that cannot be considered a catastrophe? True, but it is proof of a less than perfect movement, an upsetting thought to the passengers aboard the spacecraft. Of far more significance is the fact that at times both wobbles are synchronized. Normally the Chandler Wobble and the Annual Wobble are out of sync, and the effect on the earth's spin axis is minimal. But every seven years the two wobbles are in harmony with each other, and the wobbling of the poles becomes extreme, zigging and zagging as much as 72 feet.

In 1971 this polar drift and its peak motion every seven years were noticed by Dr. Charles A. Whitten, the chief geodesist at the National Oceanic and Atmospheric Administration. On a hunch, Whitten checked the records of earthquake activity for the previous seventy years and found another interesting and frightening correlation. The seven-year wobble cycle seemed to be directly linked to a seven-year major earthquake cycle. With very few exceptions major earthquakes have occurred when the Chandler Wobble and the Annual Wobble combined to rock the earth to the greatest degree.

Whitten warned that 1971 would be a bumper year for earthquakes because the last time the wobbles had been synchronized in 1964, the great Alaska earthquake was felt across an area of almost 500,000 square miles and released twice as much energy as the legendary San Francisco quake.

Even as the scientific journals were reporting Whitten's findings and warnings, ten severe quakes jolted the Solomon Islands in the South Pacific, while other earthquakes of major proportions ripped through New Guinea, Turkey and Chile. If any doubts remained, they were swiftly removed by the brief but savage quake that tore into the San Fernando Valley just north of Los Angeles and did a billion dollars' worth of damage and killed what officials considered a mercifully small number of only sixty-four people.

When an earthquake strikes a populated area, the death toll can grow alarmingly. In 1976 there were actually fewer "major" earthquakes than the average, but they were awesome killers. The U.S. Geological Survey reported that in 1976 more than 695,000 people died in earthquakes and quake-related disasters. Insofar as the world has kept records of such grisly statistics, only the year 1556, when 830,000 people died in Chinese earthquakes, exceeded the quake death toll. By contrast, 1975 produced only 1350 reported deaths from earthquakes and 1974 registered only 5000.

Still unexplained is whether the wobble triggers the earthquakes or the quakes cause the wobble. Dr. Whitten believes it may work both ways. The shift may trigger earthquakes, which in turn increase the wobbling at the poles. But he pulls up short of blaming the wobble as the sole cause of earthquakes.

"Many things undoubtedly enter into it," he says, "including the shifting of the earth as strain builds up beneath the surface and possibly even the pull exerted on the earth by the moon and sun. But when you add to this the earth's wobble as it reaches its maximum, you have another apparently tremendous force which may trigger earthquakes."

To this synchronization of catastrophe Cal Tech seismologist Don Anderson adds the following possibility:

"In this report," he wrote in the October 4, 1974, issue of *Science,* "I would like to point out an interesting correlation between the length of the day, Chandler wobble amplitudes and the incidence rate of great earthquakes. In particular the large deviation in the length of day around the turn of the century correlates well with the worldwide increase in global seismic activity at the same time. Smaller peaks in the length of day and seismic activity occur in the 1830s and 1940s."

Anderson blamed the changes in the length of the day, a product of reduced rotation of the earth, on changes in the patterns of wind circulation and on changes in climate. "A major volcanic eruption can lead to climatic variations that survive periods of five years or more," he pointed out. "A great outburst of volcanic activity in the last century led to an extremely long day at the turn of the century and raised speculation among the more suggestible elements of the population that God was angry and would punish man by some great catastrophe. Others believed He was pleased with man and would reward him by adding time for those who could not fit all they had to do in 24 hours."

Neither speculation, of course, was correct, but there is a definite relationship between the lengthening of the day, and the volcanic eruptions that cause it, and an increase in earthquake activity.

"The turn-of-the-century length peak also correlates well with the interval between the great decoupling and lithospheric earthquakes in Sanriko, Japan," noted Dr. Anderson. "After a great decoupling earthquake, the lithospheric plate motions can be expected to accelerate and to trigger earthquakes in adjacent portions of the arc. On the other hand, explosive volcanism in the 1830s and 1880s apparently triggered climatic changes, particularly atmospheric circulation patterns that led to changes in length of the day, changes in the earth's rotation, global earthquake activity, increases in volcanic explosions and changes in the weather."

Add to those correlations the seven-year synchronization of the Chandler Wobble and the Annual Wobble, and you have a script for planet-wide catastrophes set for 1978. And if you really want to worry, add seven years to 1978, and you arrive at 1985. That's close enough to the moment when the planets will achieve their superalignment and the Jupiter Effect will also be operative. If all these theories are correct and their cycles coincide or overlap, catastrophe will be a weak term to describe what might happen.

But it's not as if nature's more spectacular catastrophes are totally unexpected. There are about 500 major volcanoes active at the moment, while as many as 1500 major earthquakes rock the earth each year. Fortunately most occur in sparsely populated or uninhabited regions. Many occur on the sea floor and are noticed only by seismologists, who over the years have found quakes to be of enormous aid in mapping the earth's interior and understanding its composition. For earthquakes send out several types of waves, great tides of energy that travel vast distances through the earth. Two types of earthquake wave are of special interest to geophysicists. One is a compressional or P wave that alternately pulls and pushes the material through which it travels. Like a sound wave, it can travel through solids, liquids or gases, but it moves fastest through a dense solid.

The second type of earthquake wave is called a transverse or shake wave. S waves move the earth at right angles to their direction. In other words, the earth is actually shoved sideways from the direction of the wave's advance. But more important is the fact that S waves will travel only through solids.

This information has allowed geophysicists to draw a reasonably accurate picture of the earth from its core to its crustal surface. It is a picture lacking in great detail, however, for while the distance to the center of the earth is 4000 miles, no more than the distance between New York and Berlin, our instruments have poked beneath the surface of the earth a mere 5 miles. Such tentative probings are scarcely scratches on the earth's skin, which formed as the outer molten layers cooled and grew rigid. It is, however, a very thin skin, at its thickest a mere 1/400th of the earth's diameter. By contrast the skin of a peach makes up 1/200th of the fruit's mass. Beneath this very thin skin is the earth's mantle, a thick shell of hot rock that begins at an average of only 22 miles beneath the surface and continues to a depth of some 1800 miles. Within the magma, or molten rock, temperatures may reach 2500 degrees F, and pressures at the base of the mantle reach a crushing 11,000 tons per square inch. The extreme pressures and temperatures move the magma sluggishly around the earth's outer core, a liquid ball of nickel-iron where the temperatures go as high as 4000 degrees F.

This vast furnace is believed to act as an enormous electrical generator. The heat steadily churns the liquid core of iron and nickel like the slow bubbling of boiling fudge. The liquid metals act like the spinning conductors in a dynamo, generating the electrical currents that provide the earth with its magnetic shield. This magnetic force field extends out far beyond the earth, where it is called the magnetosphere and serves to shield the planet from the life-threatening radiations borne by the solar wind which streams out from the sun.

The magnetic field generated by the earth's outer core also provides people with an easy means of navigation about the earth's surface, and mariners have used the magnetic compass for a thousand years. This despite the fact that the geographic poles of the earth and its magnetic poles do not coincide. That posed far less a problem to the ancient mariners than does the question of exactly how the magnetic field came into being in the first place. Most geologists agree on the dynamo theory as the only logical explanation, for the only other alternative, that the earth is a permanent magnet, cannot withstand the application of elementary physics. Iron loses its magnetism when heated above 932 degrees F. Thus, some other explanation for the core's magnetism had to be found since its temperature is in the neighborhood of 4000° F.

That answer lies in catastrophe—specifically earthquakes. According to John Kuo and I. J. Won of Columbia's School of Engineering, about ½ of 1 percent of the enormous energies unleashed by a large earthquake impacts on the earth's solid inner core, which is thought to be composed of 80 to 85 percent iron. The even greater pressures and temperatures of the inner core keep it solid.

The waves of earthquake energy ring the earth's inner core just like a bell and cause it to oscillate the outer liquid core. This oscillation, according to Kuo and Won, "can generate the velocity field in the fluid outer core that is required by present geomagnetic theory."

The two scientists calculated that a magnitude 8.5 earthquake, such as the one that rocked Alaska in 1964, would cause a small oscillation in the outer core that repeats every seven hours and twenty-four minutes. Such an oscillation might continue its vibrations for as long as 10,000 years. But not all earthquakes increase the degree of oscillation. If they did, we all would be walking around like punchdrunk fighters, our heads echoing with the incessant bonging of a bell-ringing earth. Rather, depending on when they occur, large-magnitude quakes either cancel or reinforce the oscillation of the core. But the geomagnetic motor was probably started and continues to run on energy supplied by earthquakes.

It is in the mantle of the earth that most of its restless energies are generated and the outer character of its crust is created. "The role of the mantle in influencing conditions at the surface is manifold," according to geologist Peter J. Wyllie. "For example, during the 4.6 billion years since the origin of the earth in the solar nebula, melting of the more fusible constituents of the mantle has produced lavas that rise to the surface and solidify, adding new rocks to the crust and giving off water vapor and other gases that go to the atmosphere and the oceans. On another scale gaseous carbon compounds that came from the mantle began the story of life on the earth when they provided the raw material for organic molecules.

"Similarly the mantle as a driving force has multiple effects. The surface of the earth is shaped by the action of the mantle, moving very slowly below the crust. Mountains rise and persist because of this movement; without it erosion would wear them down to sea level within 100 million years or so. Movements of the mantle also cause volcanic eruptions, earthquakes and continental drift."

Continental drift is the most exciting theory to occur in geology in the last century. Its author was a German explorer named Alfred Wegener. Like Sir Francis Bacon before him, Wegener noticed the remarkable jigsawlike fit between the western coastline of Africa and the eastern coastline of South America. In his *Novum Organum,* published in 1620, Bacon noted that such a fit could hardly be accidental but failed to offer any reason for it. A few years later a Frenchman named R. P. François Placet proposed a catastrophic explanation: the flood as the mechanism that separated the Old World from the New.

Other geographers, geologists and explorers down through the years offered still other explanations for the phenomenon, but it was Wegener, puzzled by the mystery of ancient climates, who developed the theory in its present form. Why, he wondered, should tropical ferns have grown in London, Paris and even Greenland, while at the same time glaciers covered Brazil and the Congo?

The reason, he concluded, was that all the continents had at one time been stuck together in one huge supercontinent he called Pangaea, Greek for "all land." The supercontinent broke apart during the last years of the Mesozoic era, about 150 million years ago. The Antarctic, Australia, India and Africa broke away first to form a semi-supercontinent called Gondwanaland. Then Africa and South America snapped apart, "like pieces of a cracked ice floe, which opened the bed for the Atlantic Ocean." Some millennia later, further breakups created the present continental masses.

"When first introduced," wrote New York *Times* science editor Walter Sullivan, "Wegener's theory was a scientific heresy not unlike the one Galileo Galilei was forced to renounce before the Inquisition: that the earth moves around the sun instead of being central to all creation. As Galileo rose stiffly from his knees after denying the truth of the proposition, so the legend goes, he mumbled under his breath, 'Eppur si muove!—But still it moves!' "

Equally moving and no less heretical to the high priests of science was Wegener's theory of moving continents. But the very shape of the landmasses was to Wegener as the earth's movement about the sun was to Galileo. "It is just as if we were to refit the torn pieces of a newspaper by matching their edges," he wrote, "then check whether or not the lines of print run smoothly across. If they do, there is nothing left but to conclude that the pieces were in fact joined this way. If only one line was available for the test, we would still have found a high probability for the accuracy of the fit, but if we have *n* lines, this probability is raised to the *nth* power."

After World War I Wegener mapped the distribution of specific types of sedimentary rock and determined the position of the poles and the equator in ancient times. Three hundred million years ago the South Pole was located just east of what is today South Africa. Ninety degrees from the Pole Wegener found ample evidence of a humid, equatorial zone. That evidence takes the form of vast coal beds that reach from the eastern coast of United States all the way to China. Within those coal beds are fossilized tropical plants identical to plant fossils found in Antarctica.

Now there is still more evidence of a kind unavailable to Wegener, for it is a product of the space age. Recent satellite photos show that India is pushing China into the Pacific at the extraordinary rate of more than one inch a year. This massive nudge may account for the peculiar pattern of the many earthquakes that rock China each year. Instead of ripping along the narrow fault line areas of the continent, China's earthquakes are spread over regions hundreds of miles wide, the result of the Indian subcontinent's muscling up against China, forcing the crust to shift along a vast plate rather than a narrow fault line.

The evidence and the larger theory of plate tectonics on which it is based fit rather neatly a Hindu legend that offers both a structure for the earth and a mechanism for earthquakes. The earth, according to the legend, is supported by elephants standing on the back of a turtle, which in turn rests upon a cobra. The turtle was the incarnation of the god Vishnu; the cobra, the symbol of water. When any of these heavily burdened creatures moved, the earth's crust shifted, causing earthquakes.

The modern theory of plate tectonics is not very different. It views the earth's rigid crust as the turtle's shell, broken into six or eight huge plates, with smaller plates filling the gaps between them. Some, like the Pacific plate, are composed almost entirely of ocean floor. Other plates are part seabed and part continental landmass. North America and the western half of the North Atlantic, for example, form a single plate, stretching from San Francisco to Iceland.

The plates float about on the mantle much like congealed fat in a dish of gravy. At their edges, say, along the ocean bed where the mid-Atlantic ridge bisects the Atlantic in a north-south line, the plates are pushed apart by hot, semimolten rock rising up between the cracks in the plates. The rock boils up in great convection currents that are literally tearing the ocean floors apart at the seams. Along these seams or ridges, molten rock rises to fill the cracks as the seabed plates are carried away.

The effect of this bubbling forth of new rock is to push the plates on either side of the ridge farther and farther apart. Thus, New York and London grow farther and farther apart, but for every inch of seabed added to the Atlantic, a similar amount must be subtracted on the Pacific side of the world.

Indeed, this is the case. As the huge continental plates of North America and Eurasia are pushed apart on the Atlantic side, they are edging closer together on the Pacific. The excess seabed floor, however, is not piled up into mountains; rather, it is being dragged down the drain—literally sucked back down into the earth's interior through the vast Pacific Ocean trenches. These great rifts in the ocean bed would be, were they not covered with water, far more astonishing than the most massive mountains. The

Marianas Trench, for example, is a great slash in the earth, reaching seven miles below sea level. Many trenches are five miles and more deep and run along the floor of the Pacific for hundreds of miles.

This idea of dragging back into the earth the rock that had upwelled on the other side of the world is known more formally as the sea-floor spreading hypothesis. It was the creation of Dr. Harry Hess, a professor of geology at Princeton University, who in 1960 described the theory and announced that it "is not exactly the same as continental drift. The continents do not plow through crust impelled by unknown forces; rather they ride passively on mantle material as it comes to the surface at the crest of the ridge and then moves laterally away from it."

Then, in the inevitable action-reaction response, the sea floor is pushed down the drain of the trenches on its way deep into the earth's interior. The plate that is pulled below grinds against the adjoining plate that remains above. The grinding is one way to cause an earthquake. Another way for plates to move and jolt the earth with quakes and tremors is by sliding against each other.

This is precisely what happened in the quake that leveled Guatemala in February 1976, killing more than 20,000 people. Guatemala lies along the Motagua fault, part of the boundary line that separates the North American plate from the Caribbean plate to the south. Based on the preliminary data from seismic stations around the area, the quake and its aftershocks appear to have been caused by a series of violent lateral movements along the Motagua fault. In other words, the North American plate and the Caribbean plate rubbed edges, and the resulting earthquake killed 20,000 people.

That same rubbing of plate edges is also taking place along the borders of the Pacific and North American plates. From Baja California to San Francisco the rim of the North American plate has been ripped free of the rest of the continent by the Pacific plate. As the Pacific plate pulls upon California, it causes earthquakes along the San Andreas fault—which is the boundary line between the rim and the plate proper. The process is an ongoing one equal in its severity to the bumping China is taking at the hands of the Indian subcontinent. For those keeping their road maps accurate, the sliding push imparted by the Pacific plate will bring Los Angeles to San Francisco in about 20 million years. Of course, by then San Francisco will have moved elsewhere.

It must be obvious by now that such huge plates cannot float around forever without some large risk of collision on this cluttered earth, and that is precisely what happens.

Our present landscape is the product of a relatively recent crackup. Some 13 million years ago India crashed into Asia. Like some monstrous ocean liner gone out of control, the bow of India struck the 1500-mile beam of Asia. The two continents crunched together, crackling, splintering and heaving. From the incredible forces generated by the slow-motion collision, enormous pressures thrust jagged pinnacles of rock miles into the air. The Himalayas were born in the fury of a continental collision and are still thrusting upward. In similar epic catastrophes, the Alps, the Urals and the Appalachian Mountains were born.

The theory of continental drift has transformed not only geology, but several other scientific disciplines, including paleontology, the study of ancient, fossilized bones. Most paleontologists had explained the distribution of various mammalian animal species and the reptiles that preceded them throughout the world with land bridges: the Panamanian isthmus between the two Americas; the Bering bridge now under the chilly waters of the Bering Strait which connected North America with Asia. Australia was thought to have had a bridge joining it to Asia, while New Zealand was presumed to have been colonized by animals that hitchhiked there from Australia on serendipitously drifting logs.

But neither land bridges nor floating rafts nor Mark Spitz-style swimmers could explain such mysteries of animal distribution as the remarkable similarity between the fossils of reptiles that lived in Texas 300 million years ago and those living in Czechoslovakia some 7000 miles away. Nor could it explain away a bone found in the Transantarctic Mountains 400 miles from the South Pole. In 1967 a New Zealand geologist named Peter Barret discovered a small fossil from the jawbone of an amphibian known as a labyrinthodont.

The jawbone proved to be a catastrophic blow to the long-cherished theories of land bridges and serendipitous drifting of animals from one piece of land to another.

"Here was a fossil of great significance," declared Edwin Colbert, one of the world's leading fossil experts. "Here was some slight indication that in the distant past, Antarctica had been inhabited by land-living vertebrates.

"Students of land-living vertebrates largely ignored one continent—the island continent of Antarctica. The absence of any true land-living vertebrates, recent or extinct, on the Antarctic continent placed this great land mass, half again as large as the continental United States, generally outside the calculations of most students concerned with the distributions af ancient and recent tetrapods—the four-footed amphibians, reptiles and mammals."

The discovery of the jawbone meant, however, that Antarctica could no longer be ignored as an ancestral home of land-dwelling vertebrates, and this in turn meant that the frozen wastes of the South Pole must once have been much warmer. It also meant they must once have also been joined to the rest of the earth's landmass.

To prove the case, Colbert and a number of colleagues set up headquarters at McMurdo Station in Antarctica in 1969 and began a two-year fossil hunt. Despite horrendous working conditions, howling winds, driving clouds of snow and breath-robbing cold, the fossil hunters got to the edge of an ice field 400 miles from the South Pole. Here they found hundreds of bones from fragments to fully articulated skeletons of a host of amphibians, reptiles and mammalianlike creatures known to paleontologists as *Lystrosaurus, Thrinaxodon* and *Procolophon.*

To Colbert the findings are almost unassailable evidence for the continental drift theory. "In the first place," he says, "the presence of a diversified *Lystrosaurus* fauna in Antarctica indicates beyond any reasonable doubt that there was a dry land connection between the present south polar continent and southern Africa. Indeed, the close resemblances between fossils in the two regions, extending down to a similarity of species among various genera, is evidence that in early Triassic time, Antarctica and southern Africa were probably integral parts of a single continent. . . .

"From this demonstration of faunal and continental relationships one proceeds to the conclusions that there was such an entity as Gondwanaland, that Gondwanaland was broken asunder, that its fragments drifted apart and that

Antarctica, once the habitat of tropical or subtropical amphibians and reptiles (and abundant plants as well), came to occupy a position in a climate quite inimical to the life that had once flourished in benign temperatures."

And so, from the viewpoint of geology and paleontology, continental drift is a fact of earth life, a catastrophe in slow motion that has helped sculpture the features of the present-day surface of the earth.

There are other pieces of evidence that indicate shifts in the position of the continents that cannot be described as slow-motion catastrophe. To account for these, one requires sudden, full-blown, planet-shaking cataclysms. Such catastrophic events might well be inherent in the plate tectonic structure that overlies the magma and transforms continental drifts into great spurts of movement with correspondingly great changes in the earth's surface.

The most obvious and significant change can be seen in the earth's magnetic field. Over the last few million years the earth's magnetic field has reversed itself at least sixteen times, shifting the North Pole to the south and vice versa. Such magnetic shifts may or may not be accompanied by calamitous events. They may reflect changes in the earth's spin axis, which would call forth disasters of unthinkable magnitude. Such changes might set the continents moving to new positions at catastrophically flank speed instead of drifting at their present leisurely pace.

Some evidence to support this theory is seen in the Great Pyramids at Giza. Built 4500 years ago, the pyramids stand on a sandy plateau only 12 miles from downtown Cairo. They are marvels of human engineering, the most massive stone structures ever built by man.

But it is the Great Pyramid built by Cheops, as the Greeks called him, that may carry the clue to a great catastrophe that occurred sometime between its completion and today. Originally Cheops' pyramid was faced with pearly limestone and capped by a golden point that shot shafts of light back at the sun. Now the pyramid is dull red, 23 million blocks of sandstone uncovered 1200 years ago as the limestone face was stripped off to build the Arab mosques and palaces of Cairo. But the Great Pyramid is still mightily impressive, standing on a 13-acre base and soaring 481 feet above the desert floor. Almost from the moment of its construction, the Great Pyramid has inspired a host of legends and tales of the supernatural, as well as a spate of numerology that, depending on one's persuasion, can be seen as evidence of visitors from outer space or the mere mathematical happenstance associated with simple geometric forms.

Erich von Däniken, the messiah of outer-space visitation, notes in his book *Chariots of the Gods?* that the height of the Great Pyramid multiplied by 1 billion is equal to the distance from the earth to the sun. In the nineteenth century the astronomer royal of Scotland, Charles Piazzi Smyth, pointed out that the area of the base of the pyramid divided by twice the height equals pi, the ratio of a circle's circumference to its diameter. He also noted that the baseline of the pyramid divided by the length of one of its stones equals 365—the number of days in the year.

Other people have drawn other ratios and mathematical relationships from pyramid measurements, but the one that is perhaps most important, certainly as an indicator of changes in the earth's spin axis, is its orientation toward true geographic north. In 1940 a British archaeologist named W. M. Flinders Petrie published the first detailed survey of the pyramids of Giza. He found that the Great Pyramid of Cheops and the smaller one of Chephren that flanks it were aligned to within four minutes (of arc) west of true north. Morever, Petrie was able to prove that when they were originally built, the Great Pyramids were virtually bang on an alignment with true north.

What shifted these incredibly massive structures from their original position to a new alignment four minutes west of north? Petrie believed it was the result of a shift in the earth's poles. The idea has been largely ignored, even unknown, save to archaeologists, but now a pair of geophysicists at Edinburgh University in Scotland and Aarhus University in Denmark have examined the Petrie theory in the light of continental drift.

"Continental drift," they noted, "can cause the direction of true north to vary with respect to the moving block. The Americas have been separating from Africa and Europe owing to the spreading of the sea floor. This movement has a hinge southwest of Iceland, and is about 5 cm per year between South America and Africa. If this causes only the latter to rotate and if the rotation is uniform, in 4500 years the pyramids would be rotated 0.1° in the observed sense."

While the direction of rotation is right, the amount is far less than the actual four-minute movement made by the pyramids away from true north. But what about the other plate upon which the Arabian peninsula rides? Would its rotation, added to that of the South American and African plates, equal four minutes of arc? Again the answer is no. "Africa and the Arabian peninsula are moving apart as if hinged near the north end of the Red Sea," G. S. Pauley of Edinburgh and N. Abrahamsen of Aarhus point out. "This suggests a rotation of the pyramids in the wrong sense, but again of magnitude far too small."

Some mechanism besides continental drift was clearly at work. "Earthquakes are a possible mechanism for a local reorientation," suggest Pauley and Abrahamsen, "but a single quake of unprecedented magnitude would be needed to move the pyramids by strain release."

From their tone the two scientists clearly dislike the earthquake idea and throw the whole question open to the speculations of the geological community. "Expert geological opinion," they plead, "would be worthwhile on this point. . . ."

But must all major geologic changes occur slowly, at uniform rates over vast eons of time and always in the distant past before man arrives on the scene? Could not a combination of events shove the continental plates upon which the civilizations of man ride while he is still aboard? Is there a confluence of catastrophe that would speed the drift of continents and shift the poles in a sudden, convulsive lurch?

Similar events are thought to have taken place on Mars, where the circular scars that surround its poles are considered evidence of radical movements by the poles. As mapped by the 1972 Mariner 9, these scars mark the ancient positions of the poles and can be interpreted as evidence of a slow drift of the poles or a rather sudden change. For the Mariner photos also showed huge volcanoes and canyons pitting the face of the planet, deeper and wider than any known on earth. These volcanic structures are evidence of a violent churning and upwelling of rock from deep within the Martian interior, a mechanism very similar to the convection cells that stir the earth's magma. Unlike the earth, however, the scars of the Martian catastrophes remain visible, for the atmosphere is so thin there is scarcely any weather to erode the surface and none of the sort of plant life needed to soften the scars of geologic catastrophe.

So it might have been on the earth, where a combination of vast upwelling produced a series of massive earthquakes and volcanoes and perhaps even tilting tides within the magma itself to send the continental plates lurching about like ships steered by drunken sailors. Pulling apart here, crashing into another there, they could have shifted the spin axis of the earth and in the process changed the alignments of those structures of man that did not completely topple.

Vulcanism has played a major role not only in shaping the surface of the earth, but also in creating its atmosphere. When the earth was very young, it was cool, solid and filled with radioactive materials. Gradually that radioactivity heated up the inner portions of the earth, turning solids into liquids and gases. Sealed off from the surface, these gases under extreme pressure poked and probed at the surface lid that trapped them. In millions of places weak points were found and vented, blowing billions of cubic feet of gases and water vapor and tons of molten rock above the earth's surface. Large enough to generate a very respectable gravitational field, the newly forming earth was able to hold onto the gases and gradually to build an atmosphere and hydrosphere that covered the surface and extended out beyond the solid crust to an altitude of some 200 miles.

While geologists acknowledge the part played by volcanoes in shaping the early earth, they consider volcanic activity a relatively minor event today. Nor do they acknowledge the possibility that within recent geologic history, indeed within recorded time, exploding volcanoes have played a role in reshaping the face of the earth, not merely in the local areas where the most active volcanoes happen to be, but on a global scale. But that is precisely the picture being drawn from cores pulled from deep beneath the sea. And these cores, built for the most part of sedimentary mud, coral skeletons and segments of volcanic ash, tell a story of cataclysmic, explosive volcanic eruptions that shook the earth to its foundations. Violent in the extreme, these exploding volcanoes produced worldwide shudders that spread the sea floor not in slow, gentle movements, but in catastrophic spasms that sent the continental plates rocking over the magma like so many corks bouncing on the waves.

The new evidence is in the form of 320 cores taken from the world's oceans in as many sites. "The ash distribution in the deep sea sections," reported oceanographers James P. Kennett and Robert C. Thunell of the Graduate School of Oceanography at the University of Rhode Island, "which span the last 20 million years, indicates that there has been a much higher rate of explosive volcanism from both island arc and hot spot volcanoes during the last 2 million years."

The position of these Pacific island arc volcanoes corresponds to those areas of the oceans called subduction zones—where one plate plunges beneath another and is carried down and heated up at great depths in the earth's mantle.

Any increase in the production of material being pushed up through the sea floor in the Atlantic or anywhere along the plate boundaries would produce intense volcanic activity at the subduction zones and affect virtually the entire globe.

Exploding volcanoes, accompanied by surface-splitting earthquakes, could not only change the shape of the earth, but also drastically affect its weather. Increased volcanic activity, in fact, is now believed to be directly linked to the creation of an ice age.

"Our work," said Kennett, "shows that greatly increased volcanism during the last two million years closely coincides with that interval of earth history marked by major and rapidly oscillating climatic conditions related to glacial-interglacial cycles in the Northern Hemisphere."

The dust veils produced by volcanoes reduce the amount of sunlight reaching the earth. Volcanic dust also acts as nuclei around which ice crystals form, producing clouds, which further cut the amount of sunlight that reaches the earth.

Added to this is the increased amount of particulate matter, dust and grit being poured into the atmosphere by men's machines, which provide still more materials about which clouds may form. But the dust from volcanoes and man-made pollution does not prevent the long heat waves generated by the earth itself from being radiated out into space and lost, thus lowering temperatures still more. The result is a steadily dropping thermometer all over the earth and the creation of just the right sort of conditions for a new ice age.

# 10

# ICE AGES: PAST AND FUTURE

Patterns of nature, once perceived by science, tend to become concretized as dogma just as rigid as any the church ever created. Thus, uniformitarianism, the view that all processes involved in the earth's creation and evolution are, of necessity, incredibly long-lasting, virtually continuous and not subject to spasm, interruption, speedup or slowdown.

With this view in mind it was relatively easy for scientists to view the past geologic history of the earth and conclude it had undergone four ice ages in the last million years. Classical ice age theory also proposed a series of interglacials—periods between ice ages when the earth warmed up so that both Northern and Southern hemispheres experienced a temperate climate similar to the one in which we now luxuriate. Those interglacial periods of warmth lasted for 100,000 to 300,000 years, so the scientists believed, a nice long time scale that allowed for uniform changes to swing the weather pendulum from warm to cold and back again, the slow-motion time scale demanded by geological dogma.

It was also extremely comforting to the socio-politico-economic structure in force throughout the world, for the scientists blithely stated that we were a mere 12,000 years removed from the last ice age, giving us a margin of 88,000 to 228,000 years before we need fear a new ice age.

All that nice, neat and comforting theory crumbled, however, when a University of Miami geologist named Dr. Cesare Emiliani began looking at core samples of sediment dredged from the bottoms of the Atlantic, Pacific and Caribbean oceans. Using a method called isotopic analysis, he came to the conclusion that instead of the four ice ages described in the textbooks, there had been seven, with only six interglacial periods. Moreover, those interglacial periods of warmth had never lasted for more than 10,000 years. Needless to say, Emiliani's article published in the *Journal of Geology,* in 1955, was a disturbing idea to throw at the smug uniformitarians. Here they were cuddling up to a warm interglacial that was going to last at least another 88,000 years when the new evidence seemed to indicate we were already on our way out into the cold of a new ice age.

Of course, the Emiliani data and conclusions were at first dismissed as inconclusive. The isotopic assay system used to determine the temperatures of the different eras the cores represented was new and therefore discounted. "Isotopic analysis is very complicated both in theory and execution," acknowledged Emiliani, "and when my first paper came out nobody believed it."

And why should they, for Emiliani was drawing his temperature curves on the basis of oxygen isotopes locked in the calcium shells of ancient one-celled animals called foraminifera. From the different ratios of the isotopes trapped in the calcium, he was able to determine exact temperatures during the long course of time that the foraminifera had lived, died and settled on the sea bottom, where their shells became entombed in increasing layers of sediment.

Since the initial work other scientists have corroborated Emiliani's findings, and he has done additional studies, each yielding still more proof about the changing temperatures of the Quaternary period, the last million years of the earth's development. And nothing he or anyone else has found contradicts the original finding that the periods of gentle climates between ice ages are much shorter than had ever been thought.

"The intervals of temperatures as high as the present ones," Dr. Emiliani noted recently, "far from lasting

100,000 years or more, now appear to be short, wholly exceptional episodes in the environmental evolution of the Quaternary."

This means we may now be fast approaching the end of the warm interglacial and be headed for another ice age. Moreover, that ice age could be lurking just a few years down the road. A great debate now rages among climatologists about whether we are 1000 years away from a new ice age, or merely 100 years, while some even see it gripping us in less than 50 years and possibly within the next decade! The chilling thing is that while each one believes he is right, no one will totally discount the other's estimate.

Why this sudden spate of doom and gloom? Because the climate has been changing and with frightening rapidity.

"There is a very important climatic change going on right now," declares Reid Bryson, director of the Institute for Environmental Studies at the University of Wisconsin. "And it's not merely something of academic interest. It is something that, if it continues, will affect the whole human occupation of the earth—like a billion people starving. The effects are already showing up in rather drastic ways."

Evidence is found at every hand, in the duration of Arctic snow cover, animal migrations, sea-surface temperatures, even the types of microfossils now being found on the ocean floor. All these diverse bits of evidence, not very significant individually, together suggest a trend that points inexorably to a new ice age. For it seems that the relatively warm weather of the last fifty years is unusual.

"The past five decades," says Dr. James McQuigg, project scientist for the National Oceanic and Atmospheric Administration, "brought a period of unusually good favorable weather. We believe that we may be seeing the weather retreat into less favorable patterns."

McQuigg points to such obvious evidence of worsening weather as the seven-year severe drought in the sub-Sahara—the area known as the Sahel, encompassing the African states of Chad, Mali, Mauritania, Niger, Senegal and Upper Volta. Here a once-fertile belt bordering the desert has turned to dust as the monsoon rains move southward, dropping their water elsewhere. The resulting drought and crop loss each year leave hundreds of thousands dead of starvation and advanced malnutrition.

Drought also devastated the Russian grain harvest of 1972 and had almost as much impact again in 1975. In 1976 drought sharply reduced the winter wheat crop in the American Midwest. In Japan an unseasonably cold summer in 1973 produced a very poor harvest. Storms around the world dumped record quantities of water on the earth, but not in the drought areas. In the twelve-month period that began with Hurricane Agnes in June 1972, more water was dumped on the east coast of the United States than at any time in the past 1000 years. In the month of April 1974, in the most ravaging outbreak of tornadoes ever recorded, 148 twisters killed more than 300 people and caused half a billion dollars' damage in thirteen American states. The winter of 1976–1977 was the coldest in 100 years in the midwestern and eastern United States.

"Storms have been going where they usually don't," noted Dr. J. Murray Mitchell, Jr., another project scientist at NOAA. "Rain fronts are not moving as fast as usual. This produces floods in some areas, droughts in others. And we just don't know what's behind it all."

One possibility, of course, is that a new ice age is coming. In addition to the short-lived fury of storms, there are less transient events that frighten climatologists. They note a global cooling trend since 1940 that has reversed a temperature climb that began at the beginning of the century. As a consequence, Baffin Island in the Canadian Arctic now is covered with snowbanks the year round after forty years of being snow-free. In Iceland the pack ice has once again, after thirty years, become so thick it is a hazard to navigation. Warmth-loving animals like the armadillo, once found in abundance in the northern reaches of the American Midwest, are no longer there. They've gone south.

As a result of new techniques of computer modeling and far more data available from satellites, undersea cores, plant and animal fossils and other sources, climatologists can now look back on ancient climates and plot them with the same accuracy a market analyst plots the rise and fall of the Dow Jones. And over the long-term past the earth's climate has been as full of ups and downs as the stock market. But those ups and downs have been across a very short range. For the most part, the climate in the last 700,000 years has been absolutely lousy. In fact, for only 5 percent of that incredibly long span of time have global mean temperatures been as high as they are right now.

"It's perfectly obvious," says Reid Bryson, "that this has been the most abnormal period in at least a thousand years." The fact is that this abnormally good weather is at least as much responsible for the phenomenal yields in American crops during the last fifteen years as are improved technology and crop strains.

"The probability of getting another 15 consecutive years that good is about one in 10,000," says James McQuigg.

What does the future hold then? The same techniques that enable the climatologists to map ancient climates of the past also allow them to model future climates. From these computer models they have made some frightening extrapolations.

The course of climate change since the last glaciers retreated toward the poles some 17,000 years ago resembles the path of a small roller coaster. James Hays, an oceanographer and paleontologist at Columbia University, describes the post ice age climate as follows:

"As the ice melted, the earth then rapidly warmed until sometime between 3000 and 5000 B.C., when it was probably warmer then than today—in fact warmer than it had been for 100,000 years. But these balmy conditions were not long-lasting, and the climate gradually cooled until about 900 B.C., during the archaeological Iron Age. It then rebounded again, as evidenced by records of flourishing vineyards in England from 1000 to 1200 A.D. Thus the great Viking conquests were favored by slightly warmer conditions than today's. This time of mild climate and flourishing agriculture saw the spread of Christianity over northern and eastern Europe and the most active period of cathedral and abbey construction.

"The climate again cooled from the late fifteenth through early nineteenth centuries. This interval, the so-called Little Ice Age, was a refrigeration of considerable intensity, although still a far cry from a full ice age. It saw the greatest advance of mountain glaciers, and probably of ice on the polar seas, since the last great Ice Age. In some degree it touched every aspect of human life."

It was the Little Ice Age that finally snuffed out the 500-year life of the Vikings' colony in Greenland, which returned to its more familiar glacier-covered state. The Little Ice Age had by 1690 produced a serious famine in Great Britain, the product of eight years of continuous crop failure. In Scotland the diminishing harvests killed as many people as did the Black Plague, then rampant through Europe.

Sea ice pushed so far south it reached the Faeroe Islands, only 250 miles north of Great Britain. Following the edges of the ice pack, several kayak loads of Eskimos reached Scotland.

The Thames, which until then had frozen only rarely, became a favorite skating pond for Charles II and his court. In North America the rebelling colonists were able to *roll* their cannon across the Hudson River from the Battery to Staten Island, so solidly did it freeze each year during the Little Ice Age.

The coldest year in the Little Ice Age may have been 1816, still known in New England as "the year summer never came." Every month of the year was tinged with frost down as far as New York, and from the sixth to the nineteenth of June, subfreezing temperatures locked all New York and New England in an icy grip. During the same period a foot of snow fell on the Canadian city of Quebec.

During the Little Ice Age the glaciers, which had retreated during the warm centuries that had preceded it, once again began their march southward from the North Pole. In the Alps, whole villages built in what were considered safe sites were overwhelmed by glaciers during the seventeenth century and are still ice-covered today. In Norway the glaciers advanced by several kilometers a year during the middle of the eighteenth century. Wherever records exist of the Little Ice Age, they tell of crop failure, rising death rates and advancing glaciers.

The glacier, of course, is both the product of the ice age and its creator. Theories differ on precisely what causes glaciers, but most call for a basic process involving the accumulation of snow year after year. As the climate changes and the warm seasons grow shorter, melted snow refreezes, and the new crystals, called névé, which is French for hardened snow, become tough granular crystals far more compact than fresh snowflakes. New snow falls, further compacting the névé, air is forced out, and the crystals join together, becoming larger and denser. After a few more years the snowfields, still not glaciers, are a hundred or more feet deep, and the bottom layers have now compacted to the point where it is no longer snow or even névé, but true ice. A few more years of falling temperatures, and the ice layer has spread upward until it is half the mass, and a glacier is born.

When the glaciers reached a certain size and covered millions of square miles, spreading out over Greenland, Canada, Norway and Sweden, they began to move. Where the earth sloped, gravity pulled the monstrously huge ice sheets southward. As they approached warmer temperatures, the leading edges melted, and the ice beneath, on which the glaciers rode, also warmed. The warmer ice sped up the movement, and the glaciers at first sagged into the wells of valleys and canyons, then onto broad flatland to join other tongues of ice flowing in from neighboring basins.

What gravity started, some force within the glacier itself continues. For the glacier does not stop when it reaches the broad level plains of the lowlands. Unlike a snowball rolling down a hill, the glacier does not travel as an independent unit. It remains attached to the source of its nourishment and moves more like pancake batter spreading on a griddle. As long as the advancing glacier is fed by snows from above, it will continue to grow and move.

During the last great ice age the infant glaciers bulged out of the mountains, rolled down the gentle slopes of the foothills and combined like rivers flowing to the sea. The heavier the glaciers grew, the better they moved.

"Ice," explains University of Wisconsin geographer Gwen Schultz, "has the peculiar property of being able to move under the pressure of its own weight, provided it is thick enough. A glacier forming on a level surface will have the shape of a broad dome, flat on top and steep along the edges. When it attains a height of 100 or 200 feet (less in warm climates, more in cold) it begins to spread out. It is plastic under stress and actually flows."

Schultz saw the glaciers building up over a period of thousands of years, "thickening about the bases of the mountains on which they originated, mounting higher, filling valleys, engulfing the summits, flooding over them, and building a smooth, featureless surface over their birthplaces. They were now mature enough to move as great sheets under their own power. Higher and broader than the mountains that had suckled them, they obtained their nourishment unassisted directly from the atmosphere. And the larger they became, the greater was the surface on which snow could fall. These glaciers now rose several thousand feet skyward and presented a long, unified front against the outer world."

During the height of the last ice age, some 17,000 years ago, that unified front covered more than 30 percent of the earth's land surface. The glaciers ground down from Canada, first as huge rafts of ice that bulldozed the forests and fields, filling valleys, climbing mountains, surging and flowing until they finally buried even the peaks of the Adirondacks and the Catskills.

From Long Island the glacial ice moved ponderously westward, plowing through central Pennsylvania, southern Ohio, Indiana and Illinois. Down the Mississippi Valley the ice pushed south until it reached the junction of the Mississippi and Ohio rivers. From here the ice spread like a giant sheet, covering Missouri, Kansas, the Dakotas, Montana and Washington. Only the tallest peaks of the Rockies managed to poke above the miles-thick ice. Europe and Asia were buried by the cold mass of the Scandinavian ice cap that spread southward into the Russian Ukraine and then west across the North Sea, to cover all but the southern coast of England.

From Antarctica the glaciers surged north, while mountain glaciers poured out of the Andes to bury much of South America. Covering the land and locking much of the world's water in their frozen grip, the glaciers made Miami Beach an inland city buffeted by snowstorms every winter. The continental shelves were uncovered, and in the short summers elephants grazed on them for thousands of years.

Once established, the mighty glaciers were able to change the earth's climate to their own advantage. Their dazzling white surfaces were like mirrors throwing the sun's rays back toward the heavens, reflecting the heat and light back before it could warm the earth.

Glaciers literally feed themselves. Moist air from the oceans drifts over the continent and bumps into the advancing edge of ice. The air rises to pass over the ice mass, expanding and cooling as it goes. The dropping temperature reduces the air's ability to hold moisture. Clouds form, grow heavier and let their moisture fall as snow upon the already immense glacier.

Often the air moving over the top of a glacier cannot release its moisture because the lowest layers are so cold that the moisture cannot rise to fall out as precipitation.

Then the glacier drops the temperature of the lower air to the dew point—that temperature at which the amount of water vapor the air holds is equal to its total water-holding capacity. At this point the air must give up some of its moisture. The colder the air, the lower the capacity, and when it has reached its dew point, condensation begins. Condensation, unlike precipitation, does not fall from the air; it merely collects on solid objects or hangs suspended as particles in the air—in the form of fog. Over ice, the condensate that forms on solid objects is called hoarfrost —white crystallized ice like that found in a refrigerator that has not been defrosted for some time. Or the freezing fog can be deposited in the form of rough white ice called rime.

The hovering clouds and fog help shield the glacier from the sun's radiation, while adding to its total icy mass. Frigid winds of hurricane velocity whirl out of the glacier's center, driving the warm air before it, dropping temperatures, blowing tons of loose snow out ahead to refrigerate the warmer regions toward which it is advancing.

A glacier on the move is a thing alive. And right now many of the glaciers are moving again. The remnants of the last ice age still cover most of Greenland, where in place the glacier is an incredible 10,000 feet thick. Smaller glaciers are poised in the high mountain valleys of the Rockies, the Himalayas, the Alps and the Andes. At the bottom of the world, at the South Pole, the Antarctic continent is one enormous glacier—5 million square miles of ice—that in places is 3 miles thick.

Glaciologists have noted new advances around the world. Since 1960 about a third of the glaciers in Switzerland have begun to advance. Many of the glaciers in the Cascade and Canadian Rocky mountains are advancing once again after a retreat during the first half of this century. "During the past few centuries glaciers in mountain regions throughout the world have advanced markedly and many have attained their maximum neoglacial position," according to glacier specialists George Denton and Stephen Porter.

Will that trend continue? Does this new advance of the glaciers mean we shall soon be catapulted into a new ice age? Experts have little doubt that ice ages are a major feature of the earth's long-term past and future. Most of the debate concerns timing. And this question of timing is precisely the argument betwen the uniformitarians and the catastrophists. Donald Patten, a lay preacher by avocation and a geographer by training, argues that the ice ages were brought about suddenly by catastrophic means.

Patten believes that the ice ages occurred swiftly and catastrophically as the result of the close passage of "a single astral body (with possible ice rings or satellites)." This "astral body had a mass of between .05 and .10 of the earth and dropped 12,000,000 cubic miles of ice on it over a period of seven to eight months or in terms of Noah's voyage one year from embarkation to debarkation."

Patten offers dozens of analyses and explanations to support his thesis and to attack the uniformitarian concept of gradual accumulation of snow and slowly changing climate to create an ice age.

According to Patten, "the ice was transported across the solar system in an eccentric or comet-like orbit. It was transported from its previous orbital location or galactic region in this first phase to the Earth-Moon region. The second phase was a deflection of the cold ice particles within the Earth-Moon system by the radiation belts (Van Allen belts) of the Earth. It is thought that the particles of ice, being electrically charged, were deflected or shunted, or redirected by the magnetic field as are charged particles during periods of sunspot activity. The particles apparently converged over the magnetic polar regions, and in converging they bumped, experiencing intracollisions, which reduced their velocity, causing them to decelerate and proceed to descend. They descended over a vast magnetic polar area and concentrated in different, but proximate locations or nodes during the various descents. Thus were formed the epicenters of the nodes of the ice mass. The super-cooled ice particles descended mostly in the higher latitudes because the Magnetic North Pole happens to be located only about 1200 miles from the Geographical pole —the Earth's axis. Had the magnetic pole been located in another latitude, the Ice Epoch activity would also have been in another latitude."

Patten's "astral body," like Velikovsky's comet, is called upon to accomplish a great deal in terms of remaking the face of the earth and, of course, altering the course of its history. Both Patten's and Velikovsky's events occurred within recorded history, Velikovsky's in 1500 B.C. and

Patten's in 2800 B.C., both ancient events by our standards, but very modern when measured by astronomic yardsticks.

And cosmic events move even more slowly than do geologic phenomena. Thus, an astronomic means of starting an ice age, recently propounded by Professor W. H. McCrea of Sussex University in England, calls for a round of ice ages to occur every 100 million years. The cause: dust falling into the sun.

This idea is not very new or even original with McCrea, but he has utilized the latest knowledge of spiral galaxies to weave a new theory not only of terrestrial glaciation, but of the creation of comets and planets. The original idea, first published in 1939 by Fred Hoyle and proposed even before that by Harlow Shapley, is for the sun in its course about the galaxy to pass through a region of space that is choked with dust. The sun's gravity will pull some of the dust toward it. As it falls, the dust gives up energy, causing the sun to burn, for a brief time, even more brightly. That additional sunshine causes additional precipitation on earth, and that extra rain and snow feed the glaciers to the point where they begin to advance.

So far so good. The only problem McCrea noted was the absence of positive geological and meteorological predictions. "The work had shown rather what could occur than what had to occur. On the astronomical side, it required cloud densities much greater than, until very recently, astronomers deemed credible. . . ."

Our knowledge of the galaxy in 1939 was obviously less complete than it is now. Then we knew that the sun required about 250 million years to make one complete turn about the center of the galaxy. In its passage across the Milky Way it crossed one of the bright star-studded spiral arms of the galaxy every 100 million years. Each star-filled arm was thought to be bordered by lanes of dust. It was while passing through those lanes that the sun blazed more brightly.

It was also believed in 1939 that the spiral arms and their associated dust lanes were unstable, that over the immensely long period of time that it took the sun to cross the galaxy from one spiral arm to another, those arms and their dust trails would have been wound about the center by differential rotation into a tight coil. That would remove the arms as an explanation for long-term events such as the cause of a terrestrial ice age.

Today we have a different view of the galaxy, a view that offers a far more stable role to the spiral arms. They are now thought to be more like ripples or shock waves that fan out through the matter of the galaxy. The bright star-filled spiral arms are merely one part of the wave. The dark dust lane is seen as a strip of extra-dense matter—a very dusty part of the wave—that borders the bright arm. The pressure of the shock wave holds the arms in a position of stability against the tendency of differential rotation to wind the arms up like the spring of an alarm clock.

With the spiral arms held rigidly outstretched, McCrea sees the sun and its planets moving through the dust lanes every 100 million years, setting in motion the brighter burning process that leads to increased precipitation and ice ages on earth.

The passage through a dust lane has other implications as well. Professor McCrea believes that the most recent voyage through the dust trails bordering the Orion arm produced the comets that whip through the solar system. Peering back to previous passages some billions of years ago, McCrea points to the possibility of the planets and the sun itself being created by the repeated movement of a vast cloud of dust and gas through the shock wave spiral pattern. At every passage the shock waves would compress the cloud still further until eventually the sun and the planets would condense out.

It is a far-reaching theory that ties together a number of events in a neat, if far-flung, knot, but the primary concern here is with ice ages, and it is to this point that McCrea is most specific. For the solar system has only just emerged from a dusty strip a mere 10,000 years ago to reach its present position on the edge of the Orion arm. Ten thousand years ago, of course, is just about the time our last ice age ended.

Some geologists have already extrapolated McCrea's thesis and assume that since we have just emerged from a dust lane, we are not due to enter the next one for another 100 million years. That being the case, the next ice age is at least another 100 million years away. Unfortunately a look at the thermometer or even poking one's head out of the window a few times during the year indicates a sharp and dramatic reversal of the benign climate we have until recently been experiencing. The theory that long-term cosmic events are responsible for ice ages may be perfectly valid, but they are not the only cause of an ice age.

A modest change in the earth's orbit, an event that takes place every few thousand years or so, might be the cause of our ice ages. That is the conclusion of an international research project called CLIMAP (Climate: Long-range Investigation Mapping and Prediction), headed by geologist James D. Hays of Columbia University's Lamont-Doherty Geological Observatory.

"We are certain now that changes in the earth's orbital geometry caused the ice ages. The evidence is so strong that other explanations must now be discarded or modified."

Hays' certainty is based on seabed cores that provide a record of past temperature changes that correspond to variations in the earth's orbit. The idea is not new. More than a century ago some scientists believed the angle at which the sun's rays struck the earth could be changed by variations in the earth's orbit. This change of incident solar radiation could, they said, create an ice age. Other planets, such as Jupiter, they claimed, exerted enough gravitational pull to cause those variations.

Then, in 1920, a Serbian mathematician named Milutin Milankovitch published a paper that set the mathematics to the theory. Milankovitch had carefully worked out the orbital changes that would be involved and offered a detailed account of their effects. What the Serbian lacked was an equally detailed chronology of the earth's climatic changes to match against his mathematics.

That chronology was uncovered by the Hays team in the sedimentary cores pulled from the Indian Ocean. They have provided an unbroken geological record of the climate changes that have taken place over the last 450,000 years, a record that coincides with the Milankovitch record of the earth's orbital variations. The last measurement, by the way, showed that a peak tilting of the axis away from the perpendicular occurred 9000 years ago. The trend is now toward minimum tilt, which foreshadows a new ice age.

While Hays does not believe a full-blown ice age will freeze the earth for another few thousand years, he is very concerned about the short-term effects of the changes that are now taking place. The early stages of a cooling trend could have "a significant impact on the length of the growing season, particularly in the northern temperate zones," he says.

The time scale of the CLIMAP study is spread over thousands of years, and these data cannot be used to predict the events of the next hundred or so years. It is conceivable that other events of a purely local rather than astronomic nature could trigger an ice age. Dust, the same culprit Sussex University's Professor McCrea blamed as the cosmic cause of our ancient ice ages, is apparently equally effective when sprinkled about in the earth's atmosphere.

Unlike cosmic dust, however, atmospheric dust does not increase the sun's radiation; it reduces the amount falling on the earth. That reduction will drop temperatures and increase the size of the earth's glaciers. Many scientists believe that given enough volcanic activity, enough dust can be generated to precipitate an ice age.

The seabed cores dug up by the *Glomar Challenger* and analyzed by University of Rhode Island oceanographers James Kennett and Robert Thunell reveal a pattern of explosive volcanic activity around the world that produced more than enough ash to trigger the last ice age. Even a single, major volcanic outburst sends enough dust into the air to veil the sun and drop temperatures for short periods of time. Benjamin Franklin may have been the first scientist to notice that effect. While he served as American ambassador to France, one of the largest volcanic explosions in historic times, Laki in Iceland, took place. Franklin drew a very clear relationship between the subsequent weather and the eruption of Laki on the island of Hekla.

"During several of the summer months of the year 1783, when the effects of the Sun's rays to heat the Earth should have been the greatest, there existed a constant fog over all Europe and a great part of North America," he wrote. "This fog was of a permanent nature; it was dry and the rays of the sun seemed to have little effect towards dissipating it, as they easily do a moist fog. . . . Of course their summer effect in heating the Earth was exceedingly diminished. Hence the surface was early frozen. Hence the first snows on it remained unmelted. . . . Hence, perhaps, the winter of 1783–1784 was more severe than any that happened for many years."

This disastrous chain of climatic events Franklin blamed on the fog which he thought was due to "the vast quantity of smoke long continuing to issue during the summer from Hekla in Iceland. . . ."

The volcano that triggered such nasty weather in Franklin's time is considered by University of California ecologist Kenneth Watt number two on the list of volcanic dust producers in recorded history. "Consider the effects of the eruption of Tambora in Sumbawa, the Dutch East Indies, in April 1815," he wrote in an article for the December 23, 1972, issue of *Saturday Review*. "This was the largest known volcanic eruption in recorded history: over the period 1811 to 1818 an estimated 220 million metric tons of fine ash were ejected into the stratosphere. Of this, 150 million metric tons were added to the stratospheric load in 1815 alone, mostly in April . . . by November the average temperature of central England had dropped 4.5° F. The following twenty-four months was one of the coldest times in English history. Specifically 1816 was one of the four coldest years in the period 1698 to 1957; the coldest July in the 259-year period was in 1816; October of 1817 was the second coldest October; May of 1817 was the third coldest May."

The effect of the Tambora explosion on the world's climate was not remarked at the time, as Franklin had done when Hekla exploded. His conclusion that the smoking dust of Iceland was producing fog in Europe and subsequently dropping temperatures was all the more brilliant in the absence of devices capable of measuring the amount of solar radiation reaching the earth. But 100 years later, when the island of Krakatoa blew up in 1883, astronomers at the French observatory in Montpellier were able to record a 20 percent decrease in solar radiation. The reduction continued for more than three years, as Krakatoa poured some 50 million metric tons of ash into the global stratosphere. Much of the heavier particles rained down again into the ocean around the island, but the finer dust formed an enormous cloud, which drifted westward. Two weeks after the explosion the cloud had drifted completely around the world, spreading ever finer, blanketing more and more of the earth in its passage. It eventually made several trips around the world, diffusing out from the equator until it covered almost the entire globe.

The effect on the next year's climate was identical to that reported by Ben Franklin. Temperatures dropped in many parts of the world, and winters were uncommonly severe. The evidence of those frigid times is recorded in the annual growth rings of old trees. The University of Arizona's Tree Ring Laboratory recently sent a team to examine a group of old California trees. The rings representing 1884, the year after the Krakatoa explosion, showed evidence of a particularly hard freeze.

Since Krakatoa there have been dozens of major volcanic eruptions spewing ash, smoke and chemical pollutants such as sulfur dioxide into the air—all of which muddy the atmosphere and reduce the amount of sunlight reaching the earth. Obviously, none of the spates of volcanic activity since Krakatoa was sufficient to start us on the cold road toward the next ice age. But they all contribute, and the future looks bleaker than the past.

Volcanic activity, for example, appears to run in cycles. Most recently vulcanologists measured two active periods—from 1950 to 1956 and from 1963 to 1970. The dormant period, between 1956 and 1963, marked the lowest concentration of ash, dust and water vapor in the stratosphere. The highest concentration was achieved in 1956, when Mount Benumiaany in Siberia blew its top and sent ash 45 kilometers high.

Needless to say, the cycles are seven years long and correspond to the seven-year synchronization of the Chandler Wobble and the Annual Wobble, when earthquake activity is at its highest. According to the seven lean years-seven fat years of volcanic activity, the next increase was due to begin in 1977, while the synchronization of wobbles occurs in 1978. The Jupiter Effect, the catchy title given to the upcoming superalignment of the planets, is slated for 1983, or one year before the close of the violent volcano cycle. The Jupiter Effect is supposed to increase sunspot activity and with it produce a radical change in the weather.

At the moment, of course, we are supposed to be in a fallow period, the wobbles of the earth have not yet synchronized, the Jupiter Effect has not yet affected us, and the volcano cycle is quiescent . . . or is it? The Smithsonian Center for Short Lived Phenomena still reports an average of fifteen major explosive volcanic eruptions each year. And there is more than a little evidence that some of these eruptions occur in remote, uninhabited parts of the world, without anyone taking any immediate notice. Monitoring stations in places such as the Caucasus mountains, Mongolia and Greenland have recorded measurable increases in dustfall since 1955. Coupled with those increases are decreases in the transparency of the atmosphere and in the amount of direct sunlight reaching the earth.

Some volcanoes are poor producers of ash, simply belching forth vast volumes of lava with attendant steam and only a modicum of dust flying into the air. Others are creative—rumbling away unnoticed, deep underwater, until suddenly their lava flows build up to the point where they poke their heads above the waves and, like Athena, springing forth fully grown from Zeus' forehead, islands burst from beneath the sea. Thus was Surtsey born in 1967 off the Icelandic coast. And in similar fashion, albeit a few million years before, the Hawaiian Islands were birthed in the hissing fury of a tremendous series of underwater volcanic eruptions.

Then there are those volcanoes whose eruptions go unseen, but whose production of particulate matter is significant. One such recent eruption formed the basis of the rather aesthetic Christmas present offered to the world by the American Institute of Physics in 1974. "You may have noticed," it breezily announced in a press release, "that sunsets in many parts of the country during the last few weeks have been rather spectacularly colorful. Scientists think they now know why. As strange as it may seem, these beautiful sunsets are probably caused by the continuing eruption of *Volcán de Fuego* in Guatemala."

The press release went on to plug an article in the January 1975 issue of *Applied Optics* announcing the discovery, by Dr. Michael P. McCormick and William H. Fuller, Jr., of NASA's Langley Research Center, of two new layers of dust in the stratosphere. Using a laser radar system called Lidar, the two scientists had detected a dust layer 6 kilometers thick at 16 km of altitude. A second layer 1.5 km thick was found at an altitude of 20.25 km.

The Langley researchers did not have an estimate for the amount of sunlight screened out by the Volcán de Fuego eruption, but they did note that the same beautiful bright red, orange and yellow displays should be visible at sunrise.

It is interesting to note that until McCormick and Fuller were inspired by the spectacular effects of volcanic eruptions on sunsets, the only people to take any notice were artists. After Krakatoa exploded, Alfred Lord Tennyson penned these lines:

Had the fierce ashes of some fiery peak
Been hurl'd so high they ranged about the globe?
For day by day, thro' many a blood red eve,
In that four hundredth summer after Christ,
The wrathful sunset glared against a cross. . . .

And before poetry there was painting. The magnificent red skies in the canvases of the great English artist J. M. W. Turner were thought to be inspired by the equally magnificent sunsets produced by Tambora and volcanic explosions in the Azores at the very beginning of the nineteenth century.

Magnificent sunsets notwithstanding, nature seems to have begun to pull the triggers that may shoot us into a new ice age. The key factor seems to be the amount of radiation reaching the earth—an amount that is sharply diminished during times of increased volcanic activity. To this natural activity we add our own profligate burning of fossil fuels and further fill the air with particulate matter. Jet aircraft alone dump some 60,000 tons of small particles into the upper atmosphere each year. That is equal to 40 percent of what volcanoes manage to spew into the air. Moreover, the jet particles are of the same size as, and chemically are very similar to, volcanic ash. This almost doubles the effectiveness of the volcanic ash screen that blocks out sunlight from the earth's surface.

In an analysis of the amount of sunlight falling upon the forty-eight contiguous United States, for the period 1950 to 1972, an 8 percent decrease was noted for the autumn. Over the course of an entire year, the average drop in sunshine was 1.3 percent for the period 1964 to 1970. The National Oceanic and Atmospheric Administration offered three possible explanations for the decrease in sunshine.

1. An increase in overall cloudiness related to a trend in climate.

2. An increase in high-altitude air traffic, producing enough contrails to enhance the extent and thickness of cirrus clouds.

3. Air pollution that dims sunlight when the sun is low, either after sunrise or before sunset.

Said Dr. James K. Angell, who did the analysis: "The relatively large decrease in percentage of possible sunshine in the industrialized Northeast supports the third hypothesis."

All this means that man's hand has been added to nature's on the trigger. And it is a heavy hand indeed, for in addition to jet aerosols, we add significant amounts of dust from windblown mechanized agricultural operations, industrial burning and the production of electrical power. The burning of billions of gallons of gasoline by autos and trucks all over the world adds to the amount of particulate matter being pumped into the air every day.

Nor can we afford to ignore the effects of the more primitive slash-and-burn methods of agriculture. This approach to farming the marginal lands about the Sahara, coupled with the drought that has stricken that sad African region, has sent massive clouds of dust into the atmosphere, where they are whirled thousands of miles across the globe. Dust from the African Sahel is now producing a form of smog over Barbados Island in the West Indies. The 1973 dust concentrations over the island were 60 percent greater than in 1972 and 300 percent greater than in 1968, the first year of the African drought.

Man's contribution to the dust and particulate matter content of the atmosphere has now surpassed that of volcanoes, according to a report by the National Academy of Sciences. The effects can be seen over cities like Washington, D.C., where the turbidity of the air has increased 57 percent in the last sixty years. On a global scale such an increase would be catastrophic. Dr. Christian Junge of the Max Planck Institute in West Germany told a 1971 symposium on man's impact on climate that a 50 percent increase in turbidity from man-made sources would reduce the earth's surface temperature by 2.5 degrees F.

Such dubious achievement is not impossible. According to systems ecologist Kenneth Watt, worldwide man-made stratospheric particulate loading could reach 15 million metric tons by the year 2000. That works out to more than seven times as much dust as we managed to pump into the air in 1970. "By about 2018," he notes, "if present trends continue, the permanent particulate load in the stratosphere would be equal to Krakatau [Krakatoa] and by about 2039 the permanent load would be equal to that produced by Tambora."

These are levels we can ill afford. "As we approach the full utilization of the water, land and air, which supply our food and receive our wastes, we are becoming increasingly dependent on the stability of the present seemingly 'normal' climate," said the National Academy of Sciences Panel on Climatic Variation. "Our vulnerability to climatic change is seen to be all the more serious when we recognize that our present climate is in fact highly *abnormal,* and that we may already be producing climatic changes as a result of our own activities. This dependence of the nation's welfare, as well as that of the international community as a whole, should serve as a warning signal that we simply cannot afford to be unprepared for either a natural or a man-made climatic catastrophe."

But that seems to be precisely where we are headed. After 4.5 billion years of nature's control of the earth, the human species has in this small but potentially catastrophic area managed to seize the initiative in triggering a climate change that could precipitate another ice age.

Nor are we content merely to muddy the atmosphere and reduce the amount of sunlight reaching the earth. Through our incredibly prodigious attempts to burn virtually everything burnable, we are adding vast quantities of carbon dioxide to the gas composition of the atmosphere. This produces a capricious and paradoxical warming effect, allowing the unflappable optimists to relax.

"Aha," they say, "the greenhouse effect of carbon dioxide, which traps the sun's heat and does not permit it to be radiated back into space, will cancel out the cooling effect of the dust. Thus, our efforts will balance each other out and produce negligible effects on climate."

Unfortunately that may not be the case. The problem, according to Dr. David M. Gates, professor of botany at the University of Michigan, "is that it is extremely difficult to prove cause and effect with a giant hydrodynamic, thermodynamic machine as complex as the earth's ecosystem of ground and atmosphere."

There are, in fact, just too many variables to be taken into consideration. Clouds, oceans, surface moisture, even cities can exert massive changes on their local climates. Even though we may understand how each individual variable affects climate on a local, regional or even global scale, we do not know how they interact with each other or the ultimate effects those interactions have on overall climate.

The problem of dust aerosols and carbon dioxide is a case in point. Despite the tremendous increase in our production of carbon dioxide, the world's temperature has been steadily dropping. "From 1951 to 1972," Reid Bryson points out, "the average temperature of the North Atlantic Ocean at the surface has gone down by more than one degree Centigrade [2.5° F]."

Why? Apparently the ultimate ability of carbon dioxide in the atmosphere to increase the earth's temperature is limited. Dr. Stephen Schneider of the National Center for Atmospheric Research found that after a certain increase in atmospheric carbon dioxide, temperature increases eventually level off. We put an average of 4 percent more $CO_2$ in the air each year, but Schneider notes that an eightfold increase in the carbon dioxide concentration—which he admits is unlikely—would raise temperatures less than two degrees.

Any increase in carbon dioxide has, according to one theory, its own built-in governor, for as temperatures rise, they increase ocean evaporation. This tends to build low clouds, a factor that, according to research meteorologists Syukuro Manabe and Richard Wetherald of the National Oceanic and Atmospheric Administration, tends to screen out sunlight and prevent a further temperature rise. Thus, doubling atmospheric carbon dioxide would increase the mean surface temperature by 2.4 degrees C. But an increase of only 3 percent in the low-cloud coverage would cancel out the temperature rise.

These conclusions may be less than valid. Still other researchers see the role of dust as being minimal and our current cooling trend as merely part of a natural swing in the weather cycle. Thus, the apparent serendipitous balance struck by our addition of carbon dioxide to the earth's heat budget is accounted for. But even here there is sharp disagreement. The idea of the greenhouse effect's being self-limiting is not shared by all climatologists.

"By analogy with similar events in the past, the present natural cooling will bottom out during the next decade or so," maintains Columbia University's Wallace S. Broecker. "Once this happens, the $CO_2$ effect will tend to become a significant factor and by the first decade of the next century we may experience global temperatures warmer than any in the last 1000 years."

Added to this is the effect of waste heat produced by man's industry. The production of such waste heat, according to a National Academy of Sciences report, might become the limiting factor in the next century in deciding how much energy people can use.

"By the middle of the next century the growth of energy production could raise the mean air temperature by several degrees," says Russian geophysicist M. I. Budyko. The result, according to Budyko, is that the northern polar ice "could melt completely in the middle of the next century." Melting glaciers would raise the level of the seas to the point where virtually every coastal city on earth would be under 200 feet of water. Under such circumstances an ice age would be preferable and, it seems, on balance, more likely.

The consensus among scientists is that the next glacial period will come sooner rather than later. As Reid Bryson points out, the total temperature drop since the 1940s has been about 2.7 degrees F. This may not seem terribly dramatic, but the mean ice age temperature was only about 7 degrees F cooler than it was in 1940. Thus, we have moved one-sixth of the way toward average ice age temperatures. And there remains one major trigger that may yet be pulled and catapult us all the way into a major ice age. That trigger is the great Antarctic ice sheet. Larger than all Europe and in places as much as three miles thick, the Antarctic ice sheet would, if it slipped, cover a large part of the southern Atlantic and Pacific oceans.

Ordinarily the world's oceans are a poor reflector of sunlight, but covered with the snowy shield of the massive ice sheet, the seas would hurl so much solar energy back into space that the entire atmosphere would be chilled and a new ice age begun.

But is such a slippage likely? Recent studies by the Geological Survey of Canada revealed that the mile-thick blanket of ice that covered central Canada 8200 years ago was cleared of ice in fewer than 200 years. As the geologists reconstruct the event, the Canadian ice sheet lay on land that was below sea level. As the bottom layers of ice grew warmer, the sheet began to slip into Hudson Bay, where it broke up into smaller icebergs. Soon a channel formed into which more of the main ice sheet poured, broke up and dispersed. Once the center of the bay had been cleared, the ice sheet to the west slipped into the watery gap, followed by an equally massive surge of ice from the east.

Conditions in East Antarctica are similar to those of central Canada. Much of the land is below sea level. At Byrd Station a hole drilled 7200 feet into the ice has uncovered a layer of lubricating slush. Radar probes made from low-flying aircraft have revealed the presence of numerous lakes beneath the ice surface, indicating that the bottom layers are indeed being warmed by the heat radiating up from deep within the earth. A confluence of catastrophic events—a series of earthquakes, volcanic eruptions and violent storms lashing the earth's surface, all working to tip the globe's spin axis only slightly or merely to increase its wobble—would be enough to send the monster ice sheet plunging into the Pacific and Atlantic oceans.

Even before the great white mass could hurl back enough sunshine to chill temperatures still further, huge tidal waves would roar down on almost every coastal area on earth. Hundreds of millions of lives would, of course, be lost instantly, and those who remained to pick up the pieces would find a few decades later an ice-covered world with only the lands about the equator warm enough to raise crops.

Impossible? Not at all, and the irony is we may not even give nature a chance. Precisely such a possibility was envisioned in the 1960s by a group of strategists at the Institute for Defense Analysis in Washington, D.C. They, however, did not in their wildest speculations conceive of nature taking so malevolent a course. Their fears were that some nation or terrorist group might use nuclear weapons to shake the Antarctic ice sheet loose from the land and send it plunging into the sea.

"The most immediate effect of this vast quantity of ice surging into the water, if velocities of 100 meters per day are appropriate, would be to create massive tsunamis that would completely wreck coastal regions even in the Northern Hemisphere," wrote Dr. Gordon J. MacDonald, then executive vice-president of the institute.

Should we ever attain such heights of self-destructive idiocy, we may not want to see the future ice age that probably awaits us.

# 11

# THE DEATH OF THE DINOSAUR, THE DODO, THE WHALE AND JUST ABOUT EVERYBODY ELSE

For the sake of convenience, geologists have divided the earth's history into four broad categories, or eras. Each is based on the appearance of different fossilized life forms. The oldest era, which possesses almost no fossil evidence of life, is called the Precambrian and encompasses seven-eighths of the earth's history. The life-bearing eras that follow are called Paleozoic, Greek for ancient life; Mesozoic, for middle life; and Cenozoic, for new life. Modern dating techniques have given the Paleozoic era a span of 350 million years, the Mesozoic 150 million years and the Cenozoic the last 50 million years.

Each era is further subdivided into periods, epochs and ages. The Paleozoic era begins with the smallest subdivision, the Cambrian period age—named for a site occupied by an ancient tribe in Wales. The Cambrian was the first fossil-rich age, and the older rocks with little or no fossil record that preceded it are, as we have noted, collectively known as Precambrian.

To keep our dates straight, be advised that Day One, the first day of the Precambrian era, dawned 4.5 billion years ago, when the earth was formed. It lasted for almost 4 billion years, ending only about 570 million years ago, with the dawn of the Paleozoic era.

Within the framework of the two most recent eras two major life forms are juxtaposed—man and dinosaur—into what has become the classic anachronism. In novel, movie and comic strip the scene is the same: A group of cavemen in rude skins brandish their stone-tipped spears at a monstrous, building-high beast. The dinosaur is inevitably pictured as saber-fanged, long-clawed, mountain-tall and stupidly vicious. Men are scooped up and cracked between its jaws as easily as we chew chicken legs. Ultimately the superior intellect of the caveman—expressed in grunts and furious arm waving—enables the tribe to surround the dread beast and stab its soft underbelly with their spears. The monster roars its fury, rolls on its side with a crashing thud that brings down several dozen palm trees and raises a cloud of dust; its feet wave furiously at first and then more slowly until they stop completely. The battle is over; man has triumphed over monster.

Man, of course, is a product of the Cenozoic era. His lineage as *Homo sapiens* can be traced back only about 50,000 years, while his earlier ancestors, near-men such as *Australopithecus africanus,* first appear about 5.5 million years ago. Thus, the dinosaurs and cavemen, even *Australopithecus africanus,* could never have met on the field of battle, for the legendary monsters disappeared from the earth at the end of the Mesozoic, some 65 million years ago. But no creature of the past has so captured the human imagination as has the dinosaur. And in its popular representations the great creatures that once dominated the earth are considered mistakes of nature, great, clumsy beasts that could not adapt to the changing climate that was to create the ice age. The facts, however, are quite the opposite.

"Recent research," says Harvard paleontologist Robert T. Bakker, "is rewriting the dinosaur dossier. It appears that they were more interesting creatures, better adapted to a wide range of environments and immensely more sophisticated in their bioenergetic machinery than had been thought."

The dinosaurs, in fact, may have been among nature's most exquisitely adapted creatures. After all, they were the dominant life form on earth for 135 million years and displayed a diversity of form and function that was absolutely dazzling in its multiple adaptations. Dinosaurs were reptiles, the dominant species during the Mesozoic era. The 135 million years of their life was known as the Age of Reptiles, yet, despite their antiquity, we have known of their existence for only a little more than 100 years. It was then that the great fossil bones were uncovered and recognized as something different from the ordinary run of animals. The name "dinosaur" was coined by a nineteenth-century British scholar named Richard Owen, who responded to the rush of fossil discoveries by joining the Greek words *deinos,* for "terrible," and *sauros,* for "lizard," aptly to name the astonishing creatures conjured up by the strange bones.

And they were truly astonishing—a huge family of more than 250 different types with still more being discovered every year or so. They had armor, feathers, hair, horns, duck bills, swan necks, spines, fins and wings. They came in all sizes and shapes, some as small as chickens and others the size of a five-story building. They flew, swam, walked, lumbered and ran. They ate, slept, reproduced and dominated their time as fully as the human being does today. Our most vivid images of the ancient past are concerned with dinosaurs, yet even our most wild imaginings could not have envisioned such recent finds as:

A flying pterosaur with wings that spanned 51 feet, 10 feet greater than the wingspan of the new F-15A fighter plane. Found in Texas' Big Bend National Park only recently, the wing bones of the gigantic creature serve to confuse still further the mystery of how such enormous beasts ever flew at all.

Most paleontologists have expressed doubts that the great flying dinosaurs could flap their wings hard enough to get off the ground. Instead, they pictured them as a group of reptilian hang gliders, dragging their wings to the heights of cliffs and mountains and then leaping into the air to soar about on the thermals. But the Texas pterosaur was found in a region that during its lifetime was at least 30 miles from any mountain. If Big Bend National Park was the creature's original habitat, it would mean the giant winged reptiles were far better fliers than anyone has ever suspected. The other possibility, which preserves the hang glider theory of dinosaur flight, is that the bones were washed down from the mountains to their present position by floods or some long-vanished river.

Either that or they didn't mind the 30-mile hike into the mountains.

Then there is *Diplodocus,* among the largest of all the dinosaurs. A gigantic leaf- and tree-munching herbivore, it reached lengths of 90 feet and weighed as much as 150,000 pounds. This enormous creature has found a new measure of fame as a major character in the James Michener novel *Centennial.* His description of the beast is as good as and more readable than any to be found in a textbook:

"*Diplodocus*—'double beams'—was so named because 16 of her tail vertebrae (numbers 12 through 27 behind the hips) were made with paired flanges to protect the great artery that ran along the underside of the tail. But the vertebrae had another channel topside, and it ran from the base of the head to the strongest segment of the tail. In this channel lay a powerful, thick sinew, which was anchored securely from shoulder to hip and which could be activated from either position. Thus the long neck and the sweeping tail were similar to the main elements of the mechanical crane, which, in later eons, would lift extremely heavy objects by the clever device of running a cable over a pulley and counterbalancing the whole.

"The pulley used by *Diplodocus* was the channel made by the paired flanges of the vertebrae; her cable was the powerful sinew of neck and tail; her counterbalance was the bulk of her torso. And all functioned with an almost divine simplicity."

Despite its enormous bulk and length, *Diplodocus,* the herbivore, was prey for the king of all the predators, the ravening meat eater known as *Allosaurus.* At 50 feet long, 18 feet high and weighing only eight tons the *Allosaurus* was practically a midget compared with *Diplodocus,* but it possessed awesome weaponry. Its prehensile front feet were equipped with six-inch-long claws, while its head was almost all jaw, filled with row upon row of sharp, thin-bladed, edge-serrated, three-inch teeth. The jaws of *Allosaurus* were in their own adaptational way every bit as remarkable as the double-beamed pulley construction of the *Diplodocus* backbone. Lashed down at their rear ends by enormously powerful six-inch-thick muscles, the jaws of *Allosaurus* could bite through a tree trunk. The upper skull was of thinner bone and loosely attached to the jaws so that it could give a bit to allow the monster to swallow enormous chunks of meat.

*Allosaurus* was almost always on the prowl, tiptoeing on its hind legs, leaning forward from the hips and delicately lashing its tail back and forth to counterbalance its heavy head and powerful forebody. Spotting its prey, *Allosaurus* rushed forward, opened its enormous jaw and snapped shut on the long neck of a *Diplodocus* with enough power to crunch through flesh and bone and virtually sever the neck. Then *Allosaurus* put both front feet on its victim to hold it down and began to tear enormous chunks of flesh from the body of its victim. Its chin jutted upward, and its jaw unhinged to allow the gigantic hunk of meat to slide down its huge gullet.

And so the dinosaurs ruled the world for 135 million years, until some catastrophic event wiped them from the face of the earth. Only their bones turned-to-stone remain to mark their once-awesome presence. In their disappearance lies one of the most puzzling of the earth's mysteries—the reason for the death of the dinosaurs. For at the end of the Cretaceous period, which marked the close of the Mesozoic era, all dinosaurs, large and small, plant eaters and meat eaters, swamp dwellers and mountain climbers, fliers and swimmers, suddenly and inexplicably disappeared.

Some of their reptilian cousins—crocodiles, sea turtles, snakes—and other orders of life, such as primitive mammals, lived on, but the dinosaurs and a host of other organisms, such as the small shelled creatures called ammonites and belemnites, some forms of foraminifera and others, also disappeared as living things, leaving only their fossilized remains to note the niche they once occupied on earth.

What happened? What possible catastrophe could have overtaken the ruling class of dinosaurs and all the other creatures that died out with them?

Paleontologists have come up with many explanations, some whimsical, some incredible, some based on reptilian physiology and anatomy, some on climate, some on cataclysm and some on just plain old age, to account for the great death. No single reason can possibly account for the death of so many life forms, great and small, but all died at the same time in what was by geologic standards a very short period of time.

"There was," noted Professor W. E. Swinton, the former keeper of fossil amphibians, reptiles and birds at the British

Museum (Natural History), "no one particular cause (and perhaps it has been unrealistic to expect it) but rather a complex series of circumstances based on a fundamental geological change, complicated by racial old age (phylogeronty), lack of plasticity, both physical and mental, of the dinosaurs themselves, and with, maybe, the added factors of food deficiency, climatic variation and possibly epidemic disease."

Among the single-cause champions is the somewhat tongue-in-cheek attempt to clear up the mystery by British paleontologist A. Hallam. "Every schoolboy learns about the dramatic and relatively sudden extinction of the dinosaurs at the end of the Mesozoic Era and many are the more or less ingenious hypotheses put forward to account for it. My own favorite relates the extinction to the relative decline of the gymnosperms or naked seed plants in favour of the flowering plants during the Cretaceous period. The surviving conifer and cycad representatives include many producing oils with renowned purgative properties, from which one is drawn ineluctably to the conclusion that the poor dinosaurs died of constipation!"

Less whimsical, but equally catastrophic, is the idea of a team of German paleontologists from the University of Bonn. Digging in Provence, France, they unearthed fragments of dinosaur eggs embedded in four layers of rock. The eggshells in the oldest rocks were 2 to 4 millimeters thick. In the younger rock layers the eggshell fragments were progressively thinner. Those found in the youngest rock layer were thin enough—between 1.1 and 1.4 mm thick—to have broken very easily. Broken eggs, needless to say, do not hatch out young dinosaurs.

In another site in the Corbières region of the Pyrenees, the Bonn researchers found more eggshells that were as thin as those found in Provence.

What caused the eggshell thinning? Today we could easily blame it on DDT, but at the end of the Mesozoic, DDT was presumably in short supply. The German paleontologists reasoned that both sites were isolated swamps that were rapidly drying up. A shrinking habitat would have increased the population density of the dinosaurs until a critical degree of crowding was reached. Crowd stress then triggered some hormonal changes that caused eggshell thinning.

Perhaps, but only those two sites in southern France

have turned up dinosaur eggs with thinner-than-regulation shells. As a cause of all dinosaur deaths, it must be considered highly unlikely, as is the ever-popular idea that the newly emerging mammals ate up all the dinosaur eggs before they could hatch. This theory depends on the idea that the huge creatures were too dumb to hide or otherwise to protect their eggs.

In actual fact, the emerging mammals would have exerted as much pressure on the dinosaurs as the flea does upon the elephant today. "From their appearance in the Triassic, until the end of the Cretaceous, a span of 140 million years," Professor Bakker points out, "mammals remain small and inconspicuous while all the ecological roles of large terrestrial herbivores and carnivores were monopolized by dinosaurs; mammals did not begin to radiate and produce large species until after the dinosaurs had already become extinct at the end of the Cretaceous. One is forced to conclude that dinosaurs were competitively superior to mammals as large land vertebrates. And it would be baffling if dinosaurs were 'cold-blooded.' Perhaps they were not."

It is here, on the issue of cold-bloodedness, that the major theory of what killed the dinosaurs collapses. That theory calls for a sudden and sharp drop in temperatures, one that would reduce the cold-blooded dinosaurs to such torpor they would be unable to feed, reproduce or otherwise perform those functions essential to survival. Cold-blooded animals, such as dinosaurs and modern-day reptiles, have no natural body heat. Their body temperature rises and falls in direct response to the temperature of their environment.

To aid presumably cold-blooded dinosaurs around this problem, a pair of physicists, C. D. Bramwell and P. B. Fellgett from England's Reading University, proposed a natural solar-heating panel for one dinosaur type. *Dimetrodon grandis* was a sailback monster that once roamed the dry reaches of what is now Texas. Bramwell and Fellgett, perhaps taking a leaf from modern space technology, proposed that *Dimetrodon* used its sail—in reality a huge flap of skin running the length of the dinosaur's back and supported by upright bony struts—as a solar-heating panel. The Reading University physicists calculated that by turning its sail sideways to the sun, the dinosaur could have raised its body temperature from 26 degrees to 32 degrees

C, in an hour and twenty minutes. Without the sail, they concluded, *Dimetrodon* would need a full three hours and twenty-five minutes to achieve the same rise in body temperature.

The problem for cold-blooded dinosaurs without similar natural solar-heating panels was far more difficult and would, of course, make them pushovers for the first freeze that came along. But did one?

There is no evidence of an ice age occurring at the end of the Mesozoic. In fact, the only ice age geologists can find within 60 million years of the day the last dinosaur kicked the bucket is the one that occurred only a million and a half years ago. Moreover, there is ample evidence that the ancestors of the dinosaurs, the therapsids, some of which were as large as modern-day rhinoceroses, were quite capable of surviving the frigid winters of the Great Gondwana Ice Age, which lasted almost to the end of the Paleozoic era.

But the survival of ectothermic, or cold-blooded, creatures can be threatened by something less than a full-scale ice age. A mere drop of a few degrees in temperature, provided it is sudden, will so slow down the metabolism of ectothermic creatures as to wipe them out as a species, according to the "cold kills" theory of dinosaur extinction. That sudden temperature drop has been postulated in a number of ways. The most dramatic supposition calls for a supernova to blow up virtually on the doorstep of our solar system.

No supernova we have ever seen has exploded within 100 light-years of our solar system. That is the outermost distance astronomers calculate is near enough for the cosmic, gamma and X rays of the exploding star to change the earth's environment.

Such ear-banging supernovas do occur about once every 50 million or so years, and that would fit in at about the end of the long reign of the dinosaurs. A supernova exploding close by our solar system would have a twofold effect. First, the earth would be showered with deadly gamma rays that would kill many living things immediately and cause so many mutations that the young would be unable to survive. Secondly, the X rays blasted over the earth would blow off part of the earth's atmosphere. The incredible energy of the X rays would be transformed into heat and dropped like a lid above the earth in a layer extending from 12½ to 50 miles up.

That awesome scenario is the product of Dale Russell, a dinosaur expert with the National Museums of Canada. "The resulting turbulence," explains Russell, "would probably disrupt the heat-retaining properties of the atmosphere, generate many storms of hurricane force at the earth's surface and circulate low, water-laden air into higher, drier levels. There it would freeze to form a high altitude cover of cirrus clouds, which would reflect much of the sun's heat away from the planet. The net effect . . . would be to cause surface temperatures to drop all over the world and severely tax or exterminate organisms adapted to tropical climatic conditions."

Another means of dropping temperatures calls not for extraterrestrial events on the scale of supernova explosions, but for another form of catastrophe—the sudden ability of a tiny waterborne creature to use carbon dioxide. The incredible relationship between the minuscule organisms called plankton and a changing climate was first realized in 1971, when an American oceanographic research ship, *Glomar Challenger,* completed Leg 19 of the Deep Sea Drilling Project. Among its cores was evidence of extensive volcanic activity during the Mesozoic era. That activity continued throughout the entire era, a period that began 230 million years ago and lasted until 65 million years ago.

Initially, the extensive eruptions poured vast amounts of carbon dioxide into the atmosphere. The temperature climbed, and clouds and precipitation increased, creating an environment that encouraged the development of the huge dinosaurs. But after about 70 million years some members of the vast family of plankton that fill the seas learned to extract carbon dioxide from the atmosphere and used it to manufacture limestone shells to cover and shield their tiny bodies.

The effect was to drain large quantities of carbon dioxide from the atmosphere, thus dropping temperatures. The sediment cores from Leg 19 show that the climatic change was worldwide and the amounts of carbon dioxide used to manufacture planktonic shells were enormous.

"The earth's climate," said David W. Scholl of the U.S. Geological Survey, who led the Leg 19 expedition, "could have been cooled off just enough to absolutely raise hell with the big reptiles."

But would the effect of a colder climate, not a frigid ice age, have been enough to kill off the dinosaurs if they were endothermic—that is, warm-blooded? Before we can answer that question, we must first deal with the almost heretical idea that dinosaurs were warm-blooded. This idea flies in the face of all previous paleontological thought and challenges the accepted classification of dinosaurs as Archosauria of the class Reptilia, which every student of taxonomy will tell you means they were cold-blooded.

But perhaps the time has come to reclassify the dinosaur as an endotherm, a warm-blooded animal. "What better dividing line than the invention of endothermy?" asks Harvard's Robert Bakker, the author of the controversial idea that dinosaurs were warm-blooded. "There has been no more far-reaching adaptive breakthrough, and so the transition from ectothermy to endothermy can serve to separate the land vertebrates into higher taxonomic categories."

Bakker's iconoclastic conclusions are based on anatomical and ecological evidence. For example, the amount of energy an animal must expend for locomotion increases with the amount of speed he develops. Cold-blooded animals generate far less energy than warm-blooded animals of the same weight. Bakker estimates that if dinosaurs were cold-blooded, a 20,000-pound tyrannosaur could scamper along at only 5.8 km per hour. But a good thoroughbred brings joy to the heart of a horseplayer by running at 60 km an hour. And there is fossil evidence that dinosaurs the size of *Tyrannosaurus* were capable of running just as fast as the racehorse, while some of the smaller monsters such as the eight-foot-long *hypsilophodont* were capable of generating speeds up to 80 km an hour.

Bakker also points to size as evidence of the fact that warm blood must have coursed through dinosaur bodies. Heat generation is a function of volume, while loss of heat is a function of surface area. As the size of an animal increases, its volume increases at a greater rate than does its surface area. Small endotherms, therefore, lose a greater proportion of their heat through the skin than do large endotherms. To prevent excessive heat loss, the smaller warm-blooded creatures are insulated by covers such as hair or feathers.

A number of fossils of smaller dinosaurs such as the flying pterosaurs show the animal was insulated with a dense growth of hair or hairlike feathers, thus inspiring its name *Sordus pilosus*—hairy devil. The fossils of larger dinosaurs show them to have had hairless, noninsulated skin. Large tropical endotherms, such as the elephants, similarly maintain a constant body temperature without skin insulation.

The third piece of evidence in favor of warm blood for dinosaurs has to do with food. An endotherm must eat more than an ectotherm of the same size. In a balanced ecology the ratio of predators to prey depends on whether the dominant predator is warm- or cold-blooded. Where the predator is an ectotherm, meat eaters constitute about one-quarter of the total population. In an ecology where the dominant hunter is a carnivorous endotherm, they constitute less than 4½ percent of the population. In a number of regions fossil indications show that only 2 to 3 percent of the dinosaurs there were carnivorous.

If the dinosaurs were indeed warm-blooded, and Bakker makes a very convincing case, they might just as readily have migrated to a more temperate climate. Even if they were unable to, it is very unlikely that the cooler temperatures extended over the entire earth. The tropical belt about the equator remained tropical—if not 101 degrees F in the shade, surely not much cooler and certainly not cold enough to eliminate all the dinosaurs on the face of the earth in what by geologic standards was a sudden and catastrophic extinction.

Bakker doesn't believe in any of the most often cited reasons for the death of the dinosaurs. "When the dinosaurs fell at the end of the Cretaceous," he says, "they were not a senile, moribund group that had played out its evolutionary options. Rather they were vigorous, still diversifying into new orders and producing a variety of big-brained carnivores with the highest grade of intelligence yet present on land. What caused their fall? It was not competition, because mammals did not begin to diversify until after all the dinosaur groups (except birds) had disappeared. Some geochemical and microfossil evidence suggests a moderate drop in ocean temperature at the transition from the Cretaceous to the Cenozoic, and so cold has been suggested as the reason. But the very groups that would have been most sensitive to cold, the large crocodilians, are found as far north as Saskatchewan and as far south as Argentina before and immediately after the end of the Cretaceous. A more likely reason is the draining of shallow seas on the continents and a lull in mountain-building activity in most parts of the world, which would have produced vast stretches of monotonous topography. Such geological events decrease the variety of habitats that are available to land animals, and thus increase competition. They can also cause the collapse of intricate, highly evolved ecosystems; the larger animals seem to be the more affected. At the end of the Permian similar changes had been accompanied by catastrophic extinctions among the therapsids and other land groups. Now, at the end of the Cretaceous, it was the dinosaurs that suffered a catastrophe; the mammals and birds, perhaps because they were so much smaller, found places for themselves in the changing landscape and survived."

The dinosaurs were not alone in their dying. Their extinction at the end of the Mesozoic era was accompanied by the catastrophic disappearance of one-quarter of all the known families of animals and one-half of all species of flowering plants. Yet such a monstrous toll is not considered unusual. Rather, catastrophic extinction of life forms seems to be the norm. Somewhat akin to the sick line that goes "death is nature's way of telling us to slow down," extinction is, in fact, nature's way of increasing life's diversity and adapting it to changing ecological systems.

"If we may judge from the fossil record," notes paleontologist Norman Newell of the American Museum of Natural History, "eventual extinction seems to be the lot of all organisms. Roughly 2500 families of animals with an average longevity of somewhat less than 75 million years have left a fossil record. Of these, about a third are still living. Although a few families became extinct by evolving into new families, a majority dropped out of sight without descendants."

There is really only one explanation for such unmitigated disaster—catastrophe. "Paleontologists," acknowledges Newell, "are returning to an earlier answer: natural catastrophe."

Most species losses occurred in a spectacular fashion during the Permian period, which marked the close of the Paleozoic era. It happened 220 million years ago, when nearly half the known families of animals throughout the earth were wiped out, in the greatest mass extinction the earth has known.

By the time the Permian period had ended 75 percent of the amphibian families and more than 80 percent of the reptilian families had vanished from the face of the land. The greatest havoc, however, was wreaked among the creatures of the sea. All the trilobites, all the ancient corals, all but one class of ammonites, most of the bryozoans, brachiopods and crinoids disappeared, as did the fusulinids, complex protozoa that ranged in size from the microscopic to an inch or two. For 80 million years they had filled the shallow seas that covered the world. Their shells littered the ocean floors, forming vast limestone deposits to be quarried by their ultimate descendants, the two-legged creatures that now dominate the earth.

The Permian extinctions are even more of a puzzle to scientists than are the death of the dinosaurs, for ways to account for the mass dying out are even harder to devise. "No problem in paleontology," notes Harvard geologist Stephen Jay Gould, "has attracted more attention or led to more frustration than the search for the causes of these extinctions. The catalog of proposals would fill a Manhattan telephone book and include almost all imaginable causes: mountain building of worldwide extent, shifts in sea level, subtraction of salt from the oceans, supernovae, vast influxes of cosmic radiation, pandemics, restriction of habitat, abrupt changes in climate and so on."

Professor Gould is not calling attention to the problem simply to increase the frustration of paleontologists. No, his delineation of the problem is done to call attention to what he believes is the solution.

"When we reconstruct the history of continental movements," he wrote in *Natural History* magazine, "we realize that a unique event occurred in the latest Permian: all the continents coalesced to form the single supercontinent of Pangaea. Quite simply, the consequences of this coalescence caused the great Permian extinction."

Gould reasons that the coalescence of individual landmasses into one giant continent would eliminate the many shallow seas that had surrounded them. "Make a single square out of four square graham crackers and the total periphery is reduced by half," he instructs us. Indeed, this is the case. So a sharp reduction in living space would help eliminate many marine creatures and perhaps entire species. But there is more, and it concerns the mechanics of plate tectonics. As the continental plates crashed into each other during the late Permian, they locked together into the one super landmass called Pangaea. This union also put a temporary end to sea-floor spreading. Ocean ridges, where the new sea floor was born, sank, and this pulled the remaining shallow seas away from the continental shoulders of Pangaea. The sea level during this period may have dropped so far as to expose the entire continental shelf, thereby eliminating entirely the world's shallow seas.

"As shallow seas disappeared," Gould concludes, "the ecosystem, so rich with life in pre-Permian times, simply lacked the space to support all its members. The bag became smaller and half the marbles had to be discarded."

Most species seem to be predestined for extinction. They are created and evolve by falling into hospitable ecological niches that just coincidentally fit their specific survival talents. When they are nudged from that niche, or it changes or disappears, and when the survival traits of the species no longer fit the available ecology, extinction occurs. It is almost as if each species were sitting on the rim of a spinning roulette wheel and each time it stopped, one or more numbers came up and were thrown from the wheel into the oblivion of extinction. During the Cambrian age the trilobites, the simple shellfish that filled the warm seas of the time, swiftly evolved into many different species and then died out about 500 million years ago. The jungles of lush plants that covered the earth 350 million years ago have long since turned to coal. The dinosaurs vanished 65 million years ago, while the lumbering animals known as baluchiteria—the largest land-dwelling mammal of all times—25 feet long and 18 feet high, were wiped out some 50 million years ago.

Changing climates, cosmic radiation, elimination of shallow seas may well be responsible for some species extinction. But they do not account for all the many species that have vanished from the earth. There remains one major catastrophic mechanism of species extinction that has until recently been largely ignored. During the past 76 million years the earth has undergone 171 reversals of its magnetic field. During a reversal the intensity of the magnetic field diminishes to zero and then builds up again. While the magnetic intensity is low, cosmic radiation that normally bounces off the magnetic shield breaks through to bathe the earth.

Many scientists have noticed that each magnetic reversal is accompanied by a number of species extinctions. Dr. James D. Hays of Columbia University's Lamont-Doherty Geological Observatory notes that during the last 2.5 million years, eight species of a one-celled marine creature known as Radiolaria became extinct. Six of these extinctions followed immediately on the heels of magnetic reversals. The odds of the two events correlating with each other by chance were estimated by Dr. Hays as 1 in 10,000.

Most attempts to link cause and effect—magnetic reversals to extinction—have usually focused on the loss of the magnetic shield. With cosmic rays flooding the earth, the mutation rate rises, according to one theory. Since most mutations are harmful and result in early death of the mutants, before they can reproduce, extinction could naturally follow. But increased doses of radiation would not penetrate below the surface of the oceans, for the seas offer at least as good a shield against cosmic radiation as does the earth's magnetic field at its strongest.

Moreover, some of the Radiolaria extinctions occurred in Antarctica, which, because of its shape, is not shielded by the magnetic field, even when the field is at its peak. A magnetic reversal would not therefore produce any significant increase in radiation at Antarctica, but the Radiolaria at the South Pole were nonetheless wiped out immediately after a magnetic reversal.

Thus, marine life, at the poles, on the equator or anywhere in between, is shielded by seawater and would be unlikely to be affected by even a massive increase in cosmic radiation.

Other proposals suggest that there is a direct cause-and-effect relationship between changes in the earth's magnetic field and its climate. But here it is an increase in magnetic intensity that is responsible, shielding out or reducing some forms of solar radiation and thus resulting in a colder climate.

But, argues Dr. Ian K. Crain of the Australian National University, the thermal inertia of the oceans, which may require a century for a change in atmospheric temperature to produce even a change of a few degrees in the ocean, protects marine life from sudden climatic change. The relationship between magnetic reversals and extinction then is probably much more direct.

"Mass extinctions," says Crain, "are caused directly by the deleterious effects on organisms of the reduced magnetic field during a reversal."

The effects of low magnetism have been seen on bacteria where a fifteen-fold reduction in reproduction was quickly achieved. Movement of flatworms, protozoa and mollusks is affected by lowered magnetism, and birds show dramatic changes in motor activity. Mice placed in lower than normal magnetic fields suffer drastic changes in enzyme activity, infertility and shortened life spans.

"Infertility, changes in locomotion (hence feeding), and the incalculable effects of enzymal alterations would conceivably have had a lethal effect on many species," says Dr. Crain. Magnetic fields, of course, operate with equal effectiveness on marine as well as land organisms since seawater provides no barrier against them.

How exactly does the magnetic field affect living organisms? No one is certain but biological molecules are attracted or repelled within magnetic fields and line up in certain ways when the field is operative. Another possibility is that the magnetic field affects the electromechanical interactions of the cell, such as the movement of charged ions through the membrane.

In either case, low magnetic field effects, coupled with climatic changes and even small increases in cosmic radiation, could have been disastrous for many species. "The total effect," says Dr. Crain, "could easily have been as catastrophic as the fossil record indicates."

Might humans similarly be affected by a change in the earth's magnetic field? Since we are, despite strenuous efforts to deny it, card-carrying members of the animal kingdom, the same biological consequences could befall us. The logical question then is: When can we expect the next magnetic reversal? The answer is a new source of nightmares for the worriers among us.

In the past 10 million years the earth's magnetic poles have flipflopped just about every 220,000 years. The last magnetic switch, however, took place 700,000 years ago. The implications are inescapable: Another magnetic reversal, with all its potentially catastrophic consequences, is now long overdue.

What might bring about the next magnetic reversal?

Most geologists are agreed that the field is generated by a dynamolike action in the earth's core, and there is little reason to suspect that the dynamo runs down. Rather, one current theory says the dynamo is periodically recharged by earthquakes. There is also the theory that calls on a Velikovskian-style catastrophe to account, not only for magnetic reversals, but also for the wobble of the earth on its axis, the extinction of species and the presence of tektites, small glassy spheres that are strewn about the earth in very distinctive patterns.

It was the presence of tektites that led Drs. Saeed Durrani and Hameed Khan, physicists at the University of Birmingham in England, to attempt an explanation that would embrace all the above-mentioned phenomena. A large number of tektites fell to the bottom of the sea at the same time that the magnetic field reversed itself about 900,000 years ago. Another major concentration of tektites drifted into the seabed off the Australian continent—again at the time of the last magnetic reversal 700,000 years ago.

The two physicists decided that the earth's magnetic field might be so delicately balanced that even a slight jolt would be enough to upset it. Such a blow might easily have been delivered by a comet crashing into the earth. Khan and Durrani visualize the comet's head smashing into the surface of the earth, splashing tons of molten rock up into the air in fiery fountains. The cometary matter would harden as it fell in tektitelike patterns thousands of miles from the impact site. At the same time, cometary gases—frozen ammonia and methane—would have dispersed throughout the atmosphere and oceans, drastically altering the environment. Then, in a *Götterdämmerung* finale, the mixtures of methane and air are ignited by lightning and the skies explode in cacophony of catastrophe—a fitting accompaniment to the extinction of several species.

Until the advent of man, most species that fitted reasonably well within their ecological niches got a reasonably long spin on nature's wheel. As one species died out, another, better suited to life in the changed ecology, took its place. But now the system has changed dramatically. Man has totally usurped nature as the arbiter of species extinction. To show our concern, we now draw up lists of endangered species and pass laws to protect them from our encroachments. But whenever the two components of the problem come face to face with each other, man's desires or a species habitat, it is always the species that suffers and that must retreat still farther from our ravening technology.

There are those who argue that man has always modified his environment and the earth and that its countless life forms have not only survived, but continued much the same as ever, with little or no serious damage. That, however, is not the case. Man has modified his environment almost from the moment he began to successfully compete with the other creatures on this planet, and the earth and the other life forms have suffered for it. Man's heavy hand on the wheel of species extinction was first noticed toward the end of the last great ice age. Man at that time was a stone age hunter, clustered in small family and tribal groups about the temperate parts of Europe and Asia. He hunted everything that moved and was successful, for he is still around today, his technology far removed from stone axes and clubs, his goals still the greedy use of the earth's resources to the exclusion of all other species.

During this time the larger, slow-footed animals that were his prey were swiftly hunted into extinction—the mammoth, the woolly rhinoceros, the giant deer and the musk-ox. Most of the other potential prey, evolving at the same time as stone age man's technology, were able to develop adaptations that enabled them to cope with the threat.

Also, a major landmass—North and South America—lay for the most part beneath an ice sheet but was still capable of hosting an astonishing array of large and small animal species. Only one of the major mammalian species was not represented: man. Then the waters of the Bering Sea that separated Asia from North America receded. A land bridge connected the two continents for a short period of time about 11,000 years ago. The stage was set for the invasion of the New World by savage hunters whose skills had been honed chasing the big game of Siberia. As their prey grew more wary and evasive, the hunters crossed the Arctic Circle at the top of the world, moved east over the Bering land bridge into Alaska and slipped down a narrow passage between the temporarily halted Laurentian and Cordilleran ice sheets. What then happened was nothing short of catastrophic for the creatures of North America.

"I propose," says Paul S. Martin, professor of geosciences at the University of Arizona, "that they [the hunters] spread southward explosively, briefly attaining a density sufficiently large to overkill much of their prey."

So explosive was their spread, so savage their assault, so helpless their prey that they soon wiped from the roster of the earth's life thirty-one species of land animals. Giant ground sloths and armadillos, mammoths, unique strains of camels and horses and saber-toothed cats—in all, two-thirds of all the New World's large mammals were swiftly hunted into extinction. "There can be no repetition of this," mourned paleontologist F. Bordes, "until man lands on a planet belonging to another star."

So sudden and massive was the assault that the typical piles of animal fossil bones and associated human stone tools and other cultural materials that mark the kill sites of prehistoric man in Europe and Asia were rarely found in North America. As a result, anthropologists and paleontologists have never been happy with the idea that man was responsible for the great wave of extinctions that marked the end of the ice age in North America.

"But," offers Professor Martin in defense of his thesis, "if the new human predators found inexperienced prey, the scarcity of kill sites may be explained. A rapid rate of killing would wipe out the more vulnerable prey before there was time for the animals to learn defensive behavior, and thus the hunters would not have needed to plan elaborate cliff drives or to build clever traps. Extinction would have occurred before there was opportunity for the burial of much evidence by normal geological processes. . . . Perhaps the only remarkable aspect of New World archaeology is that *any* kill sites have been found."

The original band of hunters was probably small, no more than 100 men, women and children. But what they found was a virtual paradise. The climate was milder than the frozen tundra of eastern Siberia and western Alaska. The endemic diseases of the Old World had been left behind. These two factors alone would have guaranteed a population explosion among the new immigrants. Feeding population growth rates of only 2.4 percent into a computer, Dr. James E. Mosimann, a biostatistician at the National Institutes of Health, concluded that in 293 years the original band of 100 hunters would have grown to 300,000, a force able to annihilate *93 million* animals.

According to the model Dr. Martin has developed, the population pressures forced the prehistoric hunters to advance southward, like the wave of a blitzkrieg against an enemy front, driving the game before it. "The advance of the hunters," he notes, "was determined partly by the abundance of fresh game within the front and partly by cultural limits to the rate of human migration. In a decade or less, the population of vulnerable large animals would have been severely reduced or entirely obliterated. As the fauna vanished, the front swept on, while any remaining human population would have been driven to seeking new resources."

Thus, the catastrophic destruction of thirty-one species of animal in the New World forced man to abandon his easy hunting ways and move into the more difficult, but ultimately more rewarding, technology of agriculture.

But nature is not totally helpless before the onslaught of man on its carefully balanced ecologies. As the population of stone age hunters exploded, the game grew scarcer. The men pushed forward along a front that rolled from Canada to the Gulf of Mexico in 350 years and on to Tierra del Fuego at the tip of South America within 1000 years. And then disaster.

After destroying all the big game on the two continents, man, the destroyer, was himself threatened with starvation. He was reduced to scratching the earth with sticks and overturning rocks to feed on grubs. As a result the human population swiftly declined as the major prey animals were hunted into extinction.

Apparently we have learned little in the 11,000 years since then. Despite all our lip service to conservation of animal species, the rate of extinction is accelerating. The diversity of life on earth is now at risk, not from hunting, but from far more sophisticated methods. Industrial wastes, pesticides, loss of habitat to encroaching human demands and the ravaging of the earth for its mineral treasures—all have led to an accelerating rate of species extinction.

"The loss of diversity is not merely a matter for sentimental regret," according to Dr. Ian McTaggard Cowan, president of the Pacific Science Association, "it is a direct reduction in the number of opportunities open to future generations."

When habitats are lost or denied to animals, it is not only the immediately visible larger creatures who suffer. "Wild lands," Dr. Cowan points out, "produce food, cover for other creatures, breeding sites for insects that may be essential for pollinating crops and others that keep plant pests under control. In fact, we know little or nothing about the extinctions of such important but inconspicuous inhabitants of the earth, for our attention has been focused on the larger species."

The rate of extinction among large species has risen dramatically since the Industrial Revolution, and that same pace is probably being achieved among smaller species as a result of our increasing industrialization.

No species anywhere is safe from industry's demands. Consider the plight of Abbott's boobies, a sad-faced family of seabirds found only on small Christmas Island in the Indian Ocean. The boobies pose no threat to man, nor are they even of any specific value as a food source, nor do they have any commercial value that anyone can see. Still, Abbott's boobies are doomed to extinction because the tiny, 52-square-mile island they inhabit is unfortunately composed of almost pure phosphate of lime and thus coveted by man. Phosphate is commercially valuable as fertilizer, and an Australian-New Zealand mining company is digging away at the island, hauling away millions of tons of it every year. To speed up the process of phosphate extraction, trees are knocked down and burned, while the topsoil is scraped away to denude the land. Almost 70 percent of the island has thus been stripped of nesting sites, and the rate of reproduction among the boobies has dropped sharply. The miners will probably leave in a few years, having plundered the island down to bedrock, but that will be too late for the boobies. Naturalists feel there simply will not be enough nesting sites or even livable land left for the 4000 or so boobies that now live on the island.

As industrialization spreads to the underdeveloped nations of the tropics, the process of extinction of irreplaceable species will continue at an even greater rate. "Industrialism is the peculiar product of the temperate lands," Dr. Cowan points out. "Although some tropical environments are responsive to its methods and technologies, many are not and suffer irreparable damage from their application. The spread over the world of the industrial objectives of northern peoples can be seen as a most destructive event. Almost inevitably, diversity is sacrificed to a spurious efficiency."

As a consequence, Dr. Cowan does not view the current disappearance of species as an inevitable natural process of evolution and replacement. Analyzing animal extinctions since 1600, Dr. Cowan sees "an alarming acceleration with time." The average rate in the seventeenth century was 0.06 species a year. In the eighteenth century it was 0.05 species annually, and in the nineteenth century 0.82 per year. In this, the twentieth century, species become extinct at the rate of one a year.

Lee M. Talbot, a Smithsonian Institution ecologist and the senior scientist on the President's Council on Environmental Quality, has put the problem into even more frightening perspective. "Since the time of Christ," he points out, "man has exterminated about two percent of the known species of the world's mammals. . . . However, more than half of these losses have occurred since 1900. During the past 150 years, the rate of extermination of mammals has increased 55 fold." Projecting that same rate into the future, Talbot says somewhat gloomily, "In about 30 years all of the remaining 4000 species of mammals could be gone." About the only exceptions would be man himself and those domestic animals he chooses to raise for his own purposes.

What causes extinction of species today? Dr. Cowan listed the causes of the extinction of 600 vertebrate species as follows: "44 percent to human overkill, 57 percent to habitat alteration, 27 percent to new animal hazards and 17 percent to natural forces."

A quick tally will show the causes total more than 100 percent, but before you say "aha" and put it down to the specious arithmetic of an alarmist, be advised that some extinctions were the result of multiple causes.

Nowhere on earth is the rate of extinction as rapid as the North American continent. Pushed by increasing industrialization, urbanization and the conversion of vast tracts of wilderness into farmland, the natural wildlife of the continent has dropped sharply in abundance and number of species. Gone are the great auk, Labrador duck, passenger pigeon, heath hen and Carolina parakeet among the birds; vanished are the eastern and Merrian's elk, California, Texas, and plains grizzly bear, eastern forest bison and giant sea mink among the mammals; disappeared are the San Gorgonio trout, pahranagat spine dace, thicktail chub, harelip sucker and Ash Meadows killifish among the fish. In addition, the U.S. Department of the Interior's Bureau of Sport Fisheries and Wildlife lists 109 species of American wildlife—ranging from the timber wolf and black-footed ferret to the whooping crane and southern bald eagle—in immediate danger of extinction. Throughout the world the International Union for the Conservation of Nature and Natural Resources includes 903 species of vertebrate animals on its endangered, rare and threatened species list.

In truth, it appears that no species is exempt from sudden extinction at the hands of man. The largest creatures left on earth, the giant blue whales, face almost immediate extinction, as do twelve other members of the order Cetacea. In all, there are more than 100 species of whale, and they are fast being hunted into extinction by greedy, violent men in electronically equipped whaling fleets armed with cannon-fired harpoons. The whalers roam the seas killing whales at such a prodigious rate that the larger species will be extinct within a decade. This despite the annual meetings of an International Whaling Commission that makes pious pronouncements and establishes "kill quotas" that are virtually ignored by the nations whose whaling industries are big business.

For whales provide the raw material for pet food, motor oil, paint, margarine, soap, suntan oil, hand cream, lipstick and dozens of other products. But there is not one product of the whale for which there is not a more readily available substitute. So the great whales are being hunted into extinction, and the world will be poorer for it. The blue whale, *Sibbaldus musculus,* is the largest creature known to evolutionary history. Weighing as much as twenty-five elephants and, at 85 feet, as long as the biggest dinosaur, it is marvelously adapted to its environment. The blue whale is nonetheless doomed. Since 1905 more than 325,000 blue whales, with an aggregate weight in excess of 26 million tons, have been slaughtered.

"The tragedy of the blue whale is the reflection of an even greater one," wrote Dr. George L. Small, professor of geography at the City University of New York, "that of man himself. What is the nature of a species that knowingly and without good reason exterminates another?"

Our attack is aimed not only at the largest of the earth's creatures, but at all of them, even the smallest. For the first time since the endangered species list was established in the United States, insects are being placed on it. In March 1975 the Department of the Interior announced that of the 700 kinds of butterflies in the United States, it was placing 41 on its list of threatened and endangered species. In this case the cause is loss of habitat rather than the pesticides we have been dumping into the environment for the last forty-five years or so. And while we can trace the threat of extinction of some birds, for example, to the excessive use of DDT, which leads to a thinning of the eggshell and its subsequent destruction before it can hatch, we cannot fault pesticides here.

"Destruction of habitat is the main reason for the loss of butterflies," said Dr. Paul A. Oppler, who prepared the list as head of the Entomological Section of the Office of Endangered Species. The spread of city and suburb has wiped out many of the habitats where specific butterfly species once thrived. And apparently the extinction of butterflies is only a preview of what is to come, according to Dr. Oppler, who believes that other insect species are also faced with possible extinction.

Of all the creatures to evolve on earth, insects have demonstrated the greatest ability to survive rapid changes in their environment. They are short-lived, with dozens of generations in each century. That means that useful mutations can build up much more quickly than they do in long-lived species that have only two or three generations per century. In this way the probability of a mutation's producing the traits suddenly needed to survive the conditions of a changing environment is more likely to occur.

Insects are also remarkably prolific, bearing many young at one time. This increases the likelihood that among the vast swarm of young, one or more useful mutations will occur and be transmitted to future generations, even though a drastic environmental change might wipe out the majority of the population. Those young carrying the useful mutation that do survive will breed so quickly their offspring will soon restore the population to its original size and vigor.

With this combination of survival mechanisms, insects have become the most numerous life form on earth. A fertile queen bee, for example, can produce a population of 75,000 in the space of thirteen weeks. By contrast, a pair of elephants and all their offspring will number only about 125 after fifty years.

But even these natural survival mechanisms that have proved so successful over the last 3 billion or so years are proving vulnerable to human rapacity and technology. So what? So we wipe out the malaria-carrying mosquito, that's good, isn't it? Isn't it worth the price of the extinction of a few butterfly species? Where is the catastrophe in that?

If it were a simple trade-off—butterflies such as hairstreak and a Karner blue for an *Anopheles* mosquito or an *Aedes aegypti*—there would be no hassle. But it is not a simple matter of trading one or two beneficial insect species for one or two destructive ones. We have never been able to develop that sort of selectivity. That is not to say, of course, that we should not attempt to eliminate or at least to control the depredations of destructive insects, but it should be done with methods that do not endanger the survival of other species. We are only now beginning to understand and appreciate the incredible complexity and interdependence of the ecological network that supports all life, including man. Thus, the butterfly is not merely a pretty adjunct to the summer scene, but as necessary and effective a pollinator of food crops and flowers as is the honeybee.

Of course, if we continue at the rate we are going, there may not be any plants left to pollinate. Of the 20,000 species of vegetation on the earth, 2000 are already in imminent danger of extinction at the hand of man. By paving over the land, by poisoning it with industrial wastes, mine slurries and other means, we are changing the land and removing it as a plant habitat. Tropical rain forests, which hold 10 percent of the earth's mammals, birds and plants, are being destroyed at the rate of 14 acres a minute, according to Prince Bernhard of the Netherlands, one of the world's leading conservationists.

"Neither mankind nor rare animals have any hope for the future unless we conserve the plant kingdom, the very basis of the life support system of our planet," he said.

The danger to plant life lies not only in man's outright destruction of the land, but also in his selection of one plant species over many others as food crops. The history of agriculture reveals that man has used more than 3000 plants to feed himself, notes the prince, "but today our exploding population is being fed largely by about 15 species. And in most cases we are relying on cultivated varieties with a narrow genetic base, which combine high productivity with great vulnerability to pests and shortage of fertilizers and water."

So we ignore at our peril the lessons we should have learned so well in the course of our tenure on this planet. All life forms are interdependent. By destroying even the most insignificant creature, we may unleash catastrophic explosions of microorganisms, fungi, insects and rodents upon the earth to the detriment and even possible destruction of man.

Our repeated failures fully to appreciate the significance of this interdependence may lead us to place still another species on the endangered list, one whose ability to survive has not been demonstrated for very long. Imagine the horror, shock and disbelief when we must enter upon the endangered species list the name *Homo sapiens*, man himself.

The road to self-destruction has been for man a remarkably rapid one that began probably with the beginning of modern science. It was then he discovered that beyond the small, narrow world of his own experiences and senses there existed a vast cosmos filled with forces and energies of incredible power. "Modern science," declares Nobel Prize-winning chemist Albert Szent-Györgyi, "made man the master of cosmic forces and speeds that are out of proportion to his own. This made all previous thinking and institutions antiquated and placed man on a crossroads, one of which leads to wealth and health while the other leads to self-destruction."

The speed at which we travel and communicate has increased dramatically, and our world has shrunk accordingly. Applied to weapons, this fact means that every person on the face of the earth lives under the gun and must support the cost of incredibly complex weaponry with which to defend himself. The cost of maintaining the balance of terror so often spoken of so glibly by our statesmen has brought our economy to the brink of ruin. The combination of technological advance and antiquated thinking and institutions has led us to the brink of extinction.

"Here we stand now, on our tragically shrunken globe, with our ruined economy, with these terrific weapons in our hands, fear and distrust in our hearts," cries Szent-Györgyi.

"We either adapt to the new situation, revamp our thinking and human relations, exchange our outdated ideas of glory, force, domination and exploitation for mutual understanding, respect, help and collaboration, or else perish.

"At present we are heading for extinction and who will shed tears for us? Who regrets the dinosaur?"

# 12

# THE WORLD WILL END ON . . ?

In considering the end of the world by means of fire or ice, poet Robert Frost favored fire. He was neither the first nor the last to predict the end of the world. It was and is among the most popular speculations man has ever engaged in. The idea of disaster, of complete and utter catastrophe, has fascinated people much as the hissing cobra or the rattling of the rattlesnake mesmerizes its victim. We have expressed in our literature and mythology, in our daydreams and our nightmares a not-so-secret wish for our own destruction. Such a desire has been given many names—racial suicide, collective death wish, for example—but what is truly significant is that we have reached a point where the destruction of mankind and the earth upon which it dwells is now not only technologically possible, but statistically probable.

We seem to have achieved a state, foreseen by the French philosopher-priest Pierre Teilhard de Chardin, of "mankind falling suddenly out of love with its own destiny. This disenchantment," Teilhard wrote in *The Future of Man,* "would be conceivable and indeed inevitable, if as a result of growing reflection we came to believe that our end could only be collective death in an hermetically sealed world."

There was a time, of course, when only nature could

have had either the wit or the temerity to bring about such a conclusion to the drama of man. But we are trying hard to catch up to nature's capability. An awesome percentage of the intellectual resources of the United States and the Soviet Union is devoted to destruction. Fully 25 percent of all the scientists and engineers of the two superpowers are engaged in weapons work of some sort. By contrast, less than 1/100th of 1 percent of the group are directly engaged in arms control or disarmament.

As a result, the United States now possesses a nuclear arms capability to destroy totally every major city in the Soviet Union of over 100,000 population, not once, but thirty-six times. The Russians, hampered by an inferior technology and smaller defense spending, cannot match that capability. They can wipe out every American city of 100,000 or more people a mere twelve times.

And so at last, success. We have developed, at no small cost, the ability to equal the doomsday potential of nature. To mark that auspicious status, first achieved twenty-eight years ago, our proximity to nuclear Armageddon has been tolled by a doomsday clock mounted in the editorial offices of the *Bulletin of the Atomic Scientists*. Recently the editors moved the minute hand forward to nine minutes before midnight in frightened acknowledgment that the threat of nuclear war is more pressing than ever before.

"It behooves us as people of all nations," they said, "to remember, above all, that time is running out."

To assess that assessment, a group of the nation's leading arms control experts—all five of them—was convened at a regular meeting of the Cambridge Forum in the fall of 1975. Their conclusion: Nuclear war in some form is a distinct likelihood before the end of the twentieth century. The reason: Just about everyone either has a bomb or has access to one.

"By the end of the century," MIT political scientist George Rathjens told the symposium, "there will be several thousand reactors around the world, each producing enough material to build a weapon a week."

Nor is the technology beyond the grasp of even the illiterate members of the lunatic, terrorist fringe. At just about the same time the Cambridge Forum was meeting an educational television station in Boston was offering to the entire Public Broadcasting System a how-to-do-it course in nuclear bomb building, a sort of *Sesame Street* for the world's terrorists.

This, despite the supposedly strenuous efforts of the no longer exclusive nuclear club nations to limit the spread of atomic weapons. In fact, depending on public ideas of morality, nuclear bombs might be easier to obtain than abortions. "We will not be able to regulate nuclear weapons around the world in 1999 any better than we can control the Saturday night special, heroin or pornography today," was the conclusion of Harvard political economist Thomas Schelling, one of the five Cambridge Forum members.

But aside from the scaredy-cats at the *Bulletin of the Atomic Scientists,* the idea of a nuclear war does not seem to bother very many people. A recent *Fortune* magazine poll of business executives tested their attitudes about nuclear war. About 8 percent thought nuclear weapons would be used at a low level in war sometime during the next twenty years. About .4 percent thought there would be a substantial nuclear exchange.

"We have now a period of relative public confidence that nuclear war is not imminent," comments Dr. Paul Doty, one of the Cambridge Forum five and director of Harvard's Program for Science in International Affairs. "The complacency can itself be a danger," he said. "As time increasingly separates us from the use of nuclear weapons in war and the subsequent testing of nuclear weapons in the atmosphere, we are apt to lose the vision of how absolutely catastrophic nuclear war is. But that vision is something that must not escape us."

Yet astonishingly it does, and various concerned people and institutions feel impelled to remind us that atomic Armageddon is not all that bad. In the 1960s it was Herman Kahn's book *Thinking the Unthinkable.* More recently the National Research Council published a study of the effects of a nuclear war that would expend one-half of the suspected pile of atomic weapons held by the two superpowers. The study, thought to be the first truly scientific assessment, admitted to uncertainty where its quantitative estimates were concerned. But it concluded that despite an "unimaginable holocaust" in the target countries, the immediate physical and biological consequences to noncombatant nations would be less prolonged and less severe than many had feared.

The report went on to tick off the presumed long-range effects on the noncombatant countries lest they begin rooting for the United States and the USSR to fling nuclear missiles at each other.

"Regions of radioactivity could occur far from the detonation sites," the report warned. "Major and unfavorable climatic changes might occur;

"Were the United States and Canada involved, two thirds of the grain in international commerce would disappear;

"Public health resources in the surviving nations could be severely affected; and

"The economic, social and political consequences of the resultant worldwide terror cannot be predicted."

To avoid ending on a cheerful note, the committee concluded: "Thus, although the principal findings of this report are encouraging in a sense that they indicate that Homo sapiens—but not necessarily this civilization—would survive a major nuclear exchange, this report further underscores the urgency of halting the proliferation of nuclear weapons and, as soon as possible, reducing the world's nuclear arsenal."

Despite the distinct and real threats to civilization, the warnings not only of nuclear war, but of every catastrophe man is capable of inflicting upon himself, continue to fall upon deaf ears.

"History," says University of California at Davis zoologist Kenneth E. F. Watt, "abounds with parallels of imminent disaster. Public warnings have been ignored when they were outside the range of past experience. Consequently the appropriate countermeasures were not taken. The *Titanic* and other 'unsinkable' ships that never went down; the cities built on flood plains; Pearl Harbor and other military 'surprises,' hospitals and schools destroyed with great loss of life after repeated warnings of what fire or earthquake might do: these are some examples."

In each instance the warnings of disaster were clear and unmistakable. No boy crying "wolf" can be blamed for their being ignored; rather, we are fated, as were the ancient Greeks, to ignore our Cassandras no matter how accurate their prophecies prove to be.

"There appears to be a basic human tendency to ignore warnings about such possible enormous disasters as 'unthinkable,' " says Watt in his book *The Titanic Effect*. "We must understand this tendency and guard against it. That the world could run out of energy is 'unthinkable' and consequently it is difficult to interest people in insuring that such a thing won't happen. Yet, if we examine history, an important generalization, which might be called the *Titanic* effect, can be discerned:

"THE MAGNITUDE OF DISASTERS DECREASES TO THE EXTENT THAT PEOPLE BELIEVE THAT THEY ARE POSSIBLE, AND PLAN TO PREVENT THEM, OR TO MINIMIZE THEIR EFFECTS.

"In general, it worth taking action in advance to deal with disasters. The reason is that the costs of doing so are typically inconsequential as measured against the losses that would ensue if no such action were taken."

The costs, of course, are not always readily assessable and, where the fragile ecology of the earth is concerned, almost never predictable. Pressed by burgeoning aspirations in the third world, gluttonous energy demands in the industrialized world and incredibly exploding populations throughout the world, the earth is responding in ways that neither man nor computer can predict with any degree of precision. The one certainty is catastrophe.

And it is possible not merely as the product of complex interactions between our advanced technologies and nature, but by the simpler, more elemental effects of man barely scraping by on the subsistence level.

As the earth grows ever more crowded by people being born and not dying, we are entering an age of shortages—a time when we are running out of everything. The energy crisis is just the beginning, and it is not limited to the industrialized nations. One-third of the earth's people, an estimated 1.3 billion people, rely on firewood as fuel for cooking and heating. In their search for firewood the poor people of the world are denuding Africa, Asia and Latin America of their forests. Erik P. Eckholm, a researcher at the World Watch Institute, has found that the increasing demand for firewood has caused such widespread deforestation that subsequent erosion and loss of fertility are rendering huge tracts of land useless for food production. And population growth, aided by the humanitarian gestures of the West, puts even more pressure on the ability of the earth to support the human race.

An example of just how the system leads inevitably to catastrophe was offered by population expert Garrett

Hardin. "If you go back a generation, you will find that through the import of western medicine and food, many lives were saved in Nepal a generation ago.

"As a result of that, this is what happened.

"Man does not live by bread alone. The Nepalese needed fuel to cook their food that we sent them. They needed fuel to keep warm. . . .

"What did they do? They did the only thing they could do. They cut down the trees. . . . If you ask a Nepalese how far his grandfather had to go to get wood, he would say, 'Just a step outside the door.'

"If you ask him about his father, he would say, 'Well, he had to walk for two hours.'

"As for himself, he says, 'I have to take a two-day journey to get enough wood for my purposes.'

"Now that gives you in a nutshell what has happened to Nepal's forests. When you have deforestation, then you have loss of the soil on the hillsides. This not only makes it virtually impossible for the forests to come back in historic time, but it also means that you do not have the water-holding capacity, and so the water that lands on Nepal now reaches Bangladesh in much less time than it used to and it reaches it all in a rush.

"I think," Hardin concludes, "you can probably say that the lives that we saved in Nepal a generation ago were paid for probably many times over by lives lost in Bangladesh today."

For the Western world prospects are equally bleak. Most nations are now coming face to face with a future plagued by shortages of virtually all the resources considered vital to modern industrial society. "Many face the prospect of a series of shocks of varying severity as shortages occur in one material after another, with the first real shortages perhaps only a matter of a few years away," declared the National Academy of Sciences in a 1975 report entitled *Mineral Resources and the Environment.*

Even the vast petroleum reserves of the Middle East will be gone within thirty years at the present rates of use. U.S. copper reserves, also once presumed virtually limitless, are now in short supply, and the report suggested that the technology for recovery of copper from deep-sea nodules be explored.

Finally, the ultimate shortage, food, is already a grim fact of life in most of the world. In 1975, one of the best food production years in history, the world still suffered a shortfall equivalent to the amount of food needed to feed 130 million people. And the situation can only grow worse, for droughts and bad weather are predicted for the rest of the century.

"Another bad year in the world's major grain-producing areas will mean disaster. There are essentially no food reserves anywhere in the world," said Harvard nutritionist Jean Mayer.

Just how bad is bad? What are the parameters of the disaster forecast by Professor Mayer and other experts? As you read this, at least 100 million people are in the process of slowly starving to death. A year from now at least 10 million of them will have died, and by the end of the century one-third of the earth's present population will have died of starvation.

And there is absolutely nothing anyone can do about it. Nature will apply its own solution, catastrophically high death rates, until the population of the earth is brought back down to the point where it can be supported.

The prospects for the future are indeed grim, our society threatened on every hand by man, by nature, by overpopulation, materials shortages, starvation and a host of other disastrous events.

Consider the following visions of the apocalypse foreseen by some scientists.

"If we continue to degrade our land by dumping nutrients in the wrong places," says ecologist George M. Woodwell of the Brookhaven Laboratories, "we will eventually kill off all species of fish, fowl, birds and animals that we like, while the species that we don't like will survive. Eagles, pine trees and trout will disappear. We will be left with crabgrass, rats, crows and inedible fish."

Seismologist Peter Franken of the University of Michigan foresees a major California earthquake in the next twenty years. "Any seismologist," he adds, "will in fact be surprised if one does not occur in the next ten years. And the impact would be large. Fatalities would number several tens of thousands. These are the prompt, immediate fatalities. The injured and the later fatalities would number in the hundreds of thousands."

Nuclear physicist Dr. Howard A. Wilcox has predicted that man's output of heat into the atmosphere, the result of his energy use, if permitted to increase at present growth rates, will raise the earth's temperature enough to melt the polar ice caps and flood many of the more populous areas of the earth. Wilcox may be assuming that most of the energy is dissipated in heat, though much of it goes into frozen potential energy.

Nevertheless, the possibilities for catastrophe have been so multiplied by technology that it is difficult to sort out all the potential ramifications. Suppose one were to build a nuclear power plant on top of a fault line, where the chances of an earthquake occurring are increased exponentially.

No one could be that stupid, right? Wrong. Midway between Los Angeles and San Francisco, just 12 miles outside the picturesque Pacific coast town of San Luis Obispo, the Pacific Gas & Electric Company is building two nuclear reactors. When construction began in 1968, the power plant was placed at a point 45 miles from the Rinconada fault. According to the rules set down by the Atomic Energy Commission (since replaced by the Nuclear Regulatory Commission), all nuclear power plants sited in earthquake zones must be built to withstand the most severe jolts ever recorded in the area. So the construction engineers designed the Diablo Canyon plant to withstand a quake registering 6.75 on the Richter scale. Concrete foundations were 14 feet thick; the dome over each reactor was steel-reinforced concrete 3½ feet thick.

Then, in 1971, geologists found an underwater fault only 2½ miles west of Diablo Canyon. Called the Hosgri fault, it is believed to have caused a 1927 earthquake that hit a massive 7.25 on the Richter scale. This leaves the Diablo Canyon nuclear plant with a *minus* .50 point safety margin on the Richter scale.

To reinforce the plant and bring it into line with the safety requirements of a 7.25 magnitude earthquake will cost millions of dollars. PG&E engineers say the fault is inactive and the plant is buttressed sufficiently to withstand any quake in the area. In situations like this the NRC is faced with Hobson's choice.

In the light of the recent findings that there is probably no upper limit to the shattering power earthquakes can develop, the discussion may be academic. Given the prospect of no upper limit, there can be no safety margin, and the decision must then be based on probability: Do we want to risk a nuclear catastrophe by building a nuclear plant in an earthquake zone in order to satisfy our desire for energy? If the answer is yes, then we are exercising still another of our cost/benefit decisions in favor of racial suicide, for while the chance of disaster may be small, the magnitude of the catastrophe, if it does occur, is enormous.

If a severe quake ruptured the Diablo Canyon reactors, radioactive particles would be released into the air. They would form a colorless, odorless cloud, contaminating everything it touched, blown by the winds, poisoning the land, killing some people, causing cancer in others and genetic defects in still others yet unborn.

Such a cloud is limited in the damage it can do. Ultimately it will become so dissipated that its ability to poison and kill will be removed. But we have still other horrors to inflict on our world in the name of progress. Virtually everyone has heard of the ozone layer, that blanket of air in the upper stratosphere that serves as a shield against the fearsome ultraviolet radiation of the sun. We may now be in the process of shredding that shield with continual squirts of Freon into the atmosphere and the dumping of large quantities of other ozone-destroying chemicals from the exhausts of subsonic and supersonic jets and from the vast amounts of fertilizers we pour on humus-poor soils.

The million tons of chlorofluoromethane (Freon) manufactured each year to power our underarm deodorants and other essential products out of their cans drifts up into the atmosphere and destroys ozone. Some estimates claim that the Freon already released will reduce the ozone shield from 3 to 6 percent. A 5 percent reduction is estimated to cause 8000 new cases of skin cancer a year in the United States. A further reduction of the ozone shield would have calamitous effects. Ultraviolet radiation would become so intense it would destroy crops in the fields and alter the weather to the point where agriculture would be threatened. The oxygen supply on earth would be reduced as the phytoplankton of the world's oceans were killed, and plant and animal mutations would increase at an alarming level.

A great deal of controversy surrounds this entire subject, and while the debate continues, the ozone may or may not be in the process of destruction. It is just one more catastrophic risk we are asked to take in the name of progress.

Is there any hope then? Are we witness and participant in the final slide of the human race into Armageddon? Given the headlong plunge we seem bound and determined to continue, there is not much to be optimistic about. Certainly we can no longer turn to the *deus ex machina* of the modern world—technology—to save us.

"There is no, I repeat, no conceivable technological solution to the problems we face," declares Stanford University biologist and population expert Dr. Paul Ehrlich.

Are we doomed then to fulfill the doomsday prophecies? Is this it?

Perhaps not. We have reached a point in history that might be called a knowledge crisis. We have learned to move mountains without knowing where to put them or whether or not they should even be moved in the first place. Like the Sorcerer's Apprentice, we have learned the words to turn on the machine, but we have no idea at all of how to turn it off.

In the nineteenth century scientists saw their discoveries of the laws of the universe "leading infallibly upward to an empyrean noon hour for mankind," wrote Nobel Prize-winning biologist Jacques Monod, "whereas what we see opening before us today is an abyss of darkness."

We can turn that darkness into light by understanding that our knowledge does not render us immune to catastrophe. Rather, it can more easily bring catastrophe down upon us. When we use our technology to move mountains, we are ignoring nature's reasons for placing them there in the first place. We cannot afford to ignore nature's reasons or responses. When we dam rivers and turn them into vast lakes, the weight of the water becomes so great the earth cracks and quakes beneath the strain.

We are like greedy children set loose in a candy store. We like the goodies and ignore the inevitable consequences of our gluttony. And so we have arrived, with our vast store of knowledge, at a crisis point. From the moment we discovered fire and the weapon, we have been building toward this ultimate catastrophe of too much knowledge and not enough wisdom.

"In our frustration we sometimes blame science and technology or a particular ideology for our problems," wrote Emory University biologist W. H. Murdy, in *Science* magazine, "or we wish that evolution had taken a different direction. If however, modern society were wiped out and we were to begin again with our paleolithic ancestors, cultural evolution would inevitably lead to a similar knowledge crisis, even though its course and time of development would be different. The knowledge crisis is one that every cultural species on every inhabitable planet in the universe must surmount at a point in its evolution, or become extinct. George Wald once remarked in a lecture that it took the planet earth 4.5 billion years to discover that it was 4.5 billion years old and he added: 'Having got to that point . . . have we got much longer?' "

That is a question only we can answer. Left to their own devices, nature's plans for the end of the world are not scheduled for hundreds of millions, if not billions, of years in the future. The only question then is: Do we want to wait? Or will we continue to exercise that particularly human genius to ignore the possibilities for catastrophe?

If we do, why then, the prophecy is easy:
The world will end . . . tomorrow.

# PARTIAL BIBLIOGRAPHY

## CHAPTER ONE

Aaronson, Steve. "Shadow and Substance." *The Sciences* (June 1974).

Alfvén, Hannes. "Antimatter and Cosmology." *Scientific American* (April 1977).

———. "Plasma Physics Applied to Cosmology." *Physics Today* (February 1971).

"Alfvén on Cosmic Rays, Sunspots, Antimatter." *Physics Today* (May 1972).

"Can Matter Meet Anti-Matter Peacefully?" *Science News* (May 27, 1972).

Chandrasekhar, S. "Verifying the Theory of Relativity." *Bulletin of the Atomic Scientists* (June 1975).

Davies, P. C. W. "Is the Universe Running Away with Itself?" *Nature* (October 9, 1975).

Dickinson, D. F. "Radio Astronomy Opens Wider Windows on Space." *Smithsonian* (November 1970).

Edelson, Edward. "The Outer Limits of Space." *SR/World* (October 23, 1973).

"Experimental Tests of Relativity." *Science* (July 3, 1970).

Faulkner, John. "G-Wizardry at Dallas." *Nature* (January 24, 1975).

Gunn, J. E., and B. M. Tinsley. "An Accelerating Universe." *Nature* (October 9, 1975).
"The Infinite Universe." *Time* (December 30, 1974).
Klein, Oskar. "Arguments Concerning Relativity and Cosmology." *Science* (January 29, 1971).
Krogdahl, W. W. "The Creation of the Universe." *Sky and Telescope* (March 1973).
La Brecque, Mort. "Searching for the Antiworld." *The Sciences* (July 1972).
Llewellyn-Smith, C. "The New Unified Field Theories." *New Scientist* (April 10, 1975).
MacCallum, M. "The Breakdown of Physics?" *Nature* (October 2, 1975).
Metz, W. D. "The Decline of the Hubble Constant: A New Age for the Universe." *Science* (November 10, 1972).
———. "X-Ray Astronomy: Searching for a Black Hole." *Science* (March 15, 1973).
"More Support for the Big Bang." *Physics Today* (March 1972).
"NASA Spacecraft May Have Detected Black Hole." NASA press release (November 25, 1973).
Ostriker, J. P. "The Nature of Pulsars." *Scientific American* (January 1971).
Peebles, P. J. E., and D. T. Wilkinson. "The Primeval Fireball." *Scientific American* (June 1967).
Peruo, Bruce. "Testing Dr. Einstein." *The Sciences* (September 1973).
———. "World Without End, Amen?" *The Sciences* (November 1973).
"Radio-Wave Deflection Experiments Confirm Einstein." *Physics Today* (April 1975).
Rees, Martin, R. Ruffini, and J. A. Wheeler. "Black Holes, Gravitational Waves and Cosmology." *New Scientist* (June 5, 1975).
Schmidt, M., and F. Bello. "The Evolution of Quasars." *Scientific American* (May 1971).
Schnepper, H. W., and J. P. Delvaille. "The X-Ray Sky." *Scientific American* (July 1972).
Schram, D. N., and R. V. Wagoner. "What Can Deuterium Tell Us?" *Physics Today* (December 1974).
Stecker, Floyd W. "The Role of Antimatter in Big-Bang Cosmology." *Science and Public Affairs* (January 1974).

Stewart, Ian. "The Seven Elementary Catastrophes." *New Scientist* (November 20, 1975).
Sullivan, Walter. "Experimental Finds Challenge Accepted Theories on Atomic Physics and Cause Confusion in Science." New York *Times* (April 29, 1974).
———. "Scientists Test Idea of Infinite Universe." New York *Times* (December 20, 1974).
Thomsen, Dietrick E. "New Mathematics in Field Theory." *Science News* (April 10, 1971).
———. "Gravity, the Big Bang and the Numbers Game." *Science News* (November 16, 1974).
———. "The Universe: Chaotic or Bioselective?" *Science News* (August 24, 1974).
———. "The Weak Interaction in the Universe." *Science News* (May 31, 1975).
"The Universe May Be Half Antimatter After All." *Science News* (February 12, 1972).
Westfall, Richard S. "Newton and the Fudge Factor." *Science* (February 23, 1973).
"X-Ray Astronomy and Pulsars." *Science* (March 9, 1973).

## CHAPTER TWO

Aaronson, Steve. "The Sun Also Changes." *The Sciences* (March 1974).
Brandt, John C. "The Solar Wind Blows Some Good for Astronomy." *Smithsonian* (January 1973).
Cameron, A. G. W. "The Outer Solar System." *Science* (May 18, 1973).
"Cometary Evidence of a Planet Beyond Pluto." *Science News* (May 6, 1972).
"Does Planet X Exist?" *Sky and Telescope* (January 1973).
Elson, Benjamin M. "Planetary Formation Ideas Questioned." *Aviation Week & Space Technology* (April 15, 1974).
Greenberg, R. J., C. C. Counselman, and I. I. Shapiro. "Orbit-Orbit Resonance Capture in the Solar System." *Science* (November 17, 1972).
Grossman, L. "The Most Primitive Objects in the Solar System." *Scientific American* (February 1975).
"Halley's Comet and a Hypothetical New Planet." *Sky and Telescope* (November 1972).

Howard, Robert. "Recent Solar Research." *Science* (September 29, 1972).
Jaki, S. L. "The Titius-Bode Law: A Strange Bicentenary." *Sky and Telescope* (May 1972).
"Kohoutek: Comet of the Century." *Time* (December 17, 1973).
"The Longest Voyage." *Newsweek* (December 3, 1973).
"Lunar Science: Analyzing the Apollo Legacy." *Science* (March 30, 1973).
"Mariner 10 Mercury Encounter." *Science* (July 12, 1974).
"Maybe the Sun Is Round After All." *Physics Today* (September 1974).
"Neutrino, Little Neutral One." *Science World* (March 28, 1969).
"Origin of the Moon." *Chemistry* (February 1974).
"Pioneer Findings Paint New Picture of Jupiter." NASA press release (September 10, 1974).
"Pluto—Planet or Runaway Moon." *Science World* (April 18, 1969).
Silverberg, Gerald. "The Case of the Curious Comets." *The Sciences* (May 1973).
Urey, Harold C. "The Moon and Its Origin." *Science and Public Affairs* (November 1973).

## CHAPTER THREE

Bargmann, V., and Lloyd Motz. "On the Recent Discoveries Concerning Jupiter and Venus." *Science* (December 21, 1962).
"Carbon Monoxide Found on Jupiter for First Time." NASA press release (October 14, 1975).
"Immanuel Velikovsky Reconsidered." *Pensée,* vols. 2–10 (May 1972–Winter 1974–75).
Kolodiy, George. "Velikovsky: Paradigms in Collision." *Bulletin of the Atomic Scientists* (February 1975).
Larrabee, Eric. "Scientists in Collision: Was Velikovsky Right?" *Harper's* (August 1963).
"Lunar Probes and Velikovsky's Advance Claims." *Cosmos and Chronos* (July 1971).
MacKie, Euan. "A Challenge to the Integrity of Science?" *New Scientist* (January 11, 1973).
"Mariner 10 Venus Encounter." *Science* (March 29, 1974).

Michelson, I. "Velikovsky Forum." *Science* (July 19, 1974).
"Pioneer Findings Paint New Picture of Jupiter." NASA press release (September 10, 1974).
"Pioneer 10 and Pioneer 11." *Science* (May 2, 1975).
"The Politics of Science and Dr. Velikovsky." *American Behavioral Scientist* (September 1963).
"A Scientific Approach to Velikovsky." *Yale Scientific* (April 1967).
Sullivan, Walter. "The Velikovsky Affair." New York *Times* (October 2, 1966).
Vekhsvyatskii, Sergei. "Volcanism and the History of the Planets." *Soviet Science Review* (July 1972).
Velikovsky, I. "Are the Moon's Scars Only 3000 Years Old?" New York *Times* (early city edition, July 21, 1969).
"Velikovsky." *Industrial Research* (March 1973).
"Velikovsky: AAAS Forum for a Mild Collision." *Science* (March 15, 1974).
"Velikovsky Controversy." *Chemistry* (October 1971).
"Venus Fly-by Results and Approach to Mercury." NASA press conference (March 19, 1974).

## CHAPTER FOUR

"The Bible: The Believers Gain." *Time* (December 30, 1974).
Cowen, R. C. "Cosmic Seeds of Earthly Life." *Christian Science Monitor* (December 4, 1973).
Deloria, Vine, Jr. "Myth and the Origin of Religion." *Pensée* (September 1974).
Emiliani, C., and others. "Paleoclimatological Analysis of Late Quaternary Cores from the Northeastern Gulf of Mexico." *Science* (September 26, 1975).
Etz, D. V. "Comets in the Bible." *Christianity Today* (December 21, 1973).
Hawkins, G. S. "Astro-Archaeology—The Unwritten Evidence." *Science and Public Affairs* (October 1973).
Malim, Margaret F. "Noah's Flood." *Antiquity,* vol. 5 (1931).
Mallowan, M. E. L. "Noah's Flood Reconsidered." *Iraq,* vol. 26 (1964).

Mullen, William. "The Meso-American Record." *Pensée* (September 1974).

Raikes, R. L. "The Physical Evidence for Noah's Flood." *Iraq,* vol. 28 (1966).

Rapp, George, Jr., and others. "Pumice from Thera Identified from a Greek Mainland Archaeological Excavation." *Science* (February 2, 1973).

Rensberger, Boyce. "East Africa Fossils Suggest That Man Is a Million Years Older Than He Thinks." New York *Times* (April 12, 1975).

———. "Rock Art Shows a Supernova." New York *Times* (September 10, 1975).

Simons, E. L. "The Early Relatives of Man." *Scientific American* (July 1964).

Sullivan, Walter. "Surge of Ice Sheet's Water into Mississippi Said to Support Deluge Legends." New York *Times* (September 24, 1975).

"Tablet Found at Kish Holds Part of Epic." *Field Museum News* (June 1932).

Tomkins, Calvin. "Thinking in Time." *The New Yorker* (April 22, 1974).

"Two Million Years Are Added to History of Man." *Smithsonian* (May 1970).

## CHAPTER FIVE

Baker, H. B. "Uniformitarianism and Inductive Logic." *Pan American Geologist,* vol. 69 (1938).

"Critical Years in Geology." *Science* (January 5, 1973).

Dawson, J. W. "Some Recent Discussions in Geology." *Geological Society of America Bulletin,* vol. 7 (1894).

Gould, S. J. "Catastrophes and Steady State Earth." *Natural History* (February 1975).

———. "Is Uniformitarianism Necessary?" *American Journal of Science* (March 1965).

———. "Is Uniformitarianism Useful?" *Journal of Geological Education* (October 1967).

———. "Reverend Burnet's Dirty Little Planet." *Natural History* (April 1975).

Hawkes, Leonard. "Some Aspects of the Progress in Geology in the Last Fifty Years, II." *Geological Society of London Quarterly Journal,* vol. 114 (1958).

Hutton, James. "Theory of the Earth; or an Investigation of the Laws Observable in the Composition, Dissolution and Restoration of Land Upon the Globe." *Transactions of the Royal Society of Edinburgh,* vol. 1 (1788).

Krynine, P. D. "Uniformitarianism Is a Dangerous Doctrine." *Journal of Paleontology,* vol. 30 (1956).

Rudwick, M. J. S. "Sir Charles Lyell." *Nature* (October 9, 1975).

Sutton, John. "Charles Lyell and the Liberation of Geology." *New Scientist* (February 20, 1975).

## CHAPTER SIX

"Creationism." *Scientific American* (January 1971).

"Darwin's Big Book." *Science* (May 23, 1975).

De Camp, L. S. "The End of the Monkey War." *Scientific American* (February 1969).

"Evolutionary Thinking." *Newsweek* (July 7, 1975).

Galston, A. W. "Botanist Charles Darwin." *Natural History* (December 1973).

Ghiselin, M. T. "Darwin and Evolutionary Psychology." *Science* (March 9, 1973).

Gould, S. J. "Darwin's Dilemma." *Natural History* (June–July 1974).

———. "Darwin's Delay." *Natural History* (December 1974).

Harper, C. W., Jr. "Origin of Species in Geologic Time." *Science* (October 3, 1975).

"Huxley Defends Darwin." *Scientific American* (January 1871).

Johnson, Irving and Electa. "Lost World of the Galápagos." *National Geographic* (May 1959).

Johnson, M. P., and P. H. Raven. "Species Number and Endemism: The Galápagos Archipelago Revisited." *Science* (March 2, 1973).

Muller, Herbert J. "Reflections on Re-reading Darwin." *Bulletin of the Atomic Scientists* (February 1973).

"Neutralists v. Selectionists." *Science* (May 11, 1973).

Newell, N. D. "Evolution Under Attack." *Natural History* (April 1974).

Rensberger, B. "Evolution Theory Still in Dispute Fifty Years After the Monkey Trial." New York *Times* (July 10, 1975).

———. "Sociobiology: Updating Darwin on Behavior." New York *Times* (May 28, 1975).

## CHAPTER SEVEN

"Amber Light for Genetic Manipulation." *Nature* (January 31, 1975).
Berg, Paul, and others. "Potential Biohazards of Recombinant DNA Molecules." *Science* (July 26, 1974).
Blakeslee, S. "Molecular Findings Offer Evolutionists a Challenge." New York *Times* (June 19, 1975).
Cairns-Smith, G. "Genes Made of Clay." *New Scientist* (October 24, 1974).
Danielli, J. F. "Artificial Synthesis of New Life Forms." *Bulletin of the Atomic Scientists* (December 1972).
"Genetic Engineers Discuss Our Future." *New Scientist* (March 6, 1975).
"Genetics: Conference Sets Strict Controls to Replace Moratorium." *Science* (March 14, 1975).
"Group Gives Go-Ahead to Genetic Engineering." *Chemical and Engineering News* (February 3, 1975).
Hallam, A. "Extended View of Evolution." *Nature* (May 22, 1975).
Jukes, T. H., and R. Holmquist. "Evolutionary Clock: Nonconstancy of Rate in Different Species." *Science* (August 11, 1972).
Lucas, E. C. "Creationism and Evolutionism." *Science* (March 9, 1973).
Luria, Salvador E. "Genesis." *Natural History* (June–July 1973).
McDonald, H. "Implanting Human Values into Genetic Control." *Science and Public Affairs* (February 1974).
"NAS Ban on Plasmid Engineering." *Nature* (July 19, 1974).
Prigogine, I., and others. "Thermodynamics of Evolution." *Physics Today* (November 1972; December 1972).
Rensberger, B. "Genetic Mutations Called Minor Evolutionary Factor." New York *Times* (April 18, 1975).
Thoday, J. M. "Non-Darwinian 'Evolution' and Biological Progress." *Nature* (June 26, 1975).
Vanderkooi, G. "Evolution as a Scientific Theory." *Christianity Today* (May 7, 1971).

Weinberg, J. H. "Evolution, a Theory Evolving." *Science News* (February 22, 1975).
———. "Plasmid—Welcome to the Word Pool. But What Are You?" *Science News* (June 21, 1975).

## CHAPTER EIGHT

"Amino Acids in Both Moon and Meteorite." *Physics Today* (February 1971).
Barghoorn, E. S. "The Oldest Fossils." *Scientific American* (May 1971).
Bar-Nun, A., and others. "Shock Synthesis of Amino Acids in Simulated Primitive Environments." *Science* (April 24, 1970).
Carter, Charles. "Cradles for Molecular Evolution." *New Scientist* (March 27, 1975).
"Cosmic Rays and the Origin of Life." *Chemistry* (April 1975).
Eglinton, G., and M. Calvin. "Chemical Fossils." *Scientific American* (January 1967).
"Extraterrestrial Amino Acids." *Science News* (June 26, 1971).
Fox, Sidney W. "How Did Life Begin?" *Science* (July 22, 1960).
———. "In the Beginning . . . Life Assembled Itself." *New Scientist* (February 27, 1969).
———. "Origin of the Cell: Experiments and Premises." *Naturwissenschaften,* vol. 60 (1973).
———. "Origins of Biological Information and the Genetic Code." *Molecular and Cellular Biochemistry* (April 15, 1974).
Frieden, Earl. "The Chemical Elements of Life." *Scientific American* (July 1972).
"Is the Genetic Code the Best Chemical Language?" *New Scientist* (June 5, 1975).
Kiefer, I. "Proving We Are the Stuff of Which Stars Are Made." *Smithsonian* (May 1972).
Kovacs, K. L., and A. S. Garay, "Primordial Origins of Chirality." *Nature* (April 10, 1975).
Pagel, B. "Our Origin in the Stars." *Spectrum,* vol. 6 (1975).
Ponnamperuma, C. "The Emergence of Life." *AAAS Symposium on Cosmic Evolution* (February 26, 1974).

"Protobiogenesis." *Nature* (April 20, 1973).
Sullivan, Walter. "Molecular Astronomy: The Great Void Is Alive." New York *Times* (September 28, 1975).
"Synthesis of Amino Acids from Gases Known in Space." *Science News* (November 27, 1971).
Turcotte, D. L., and others. "Evolution of the Moon's Orbit and the Origin of Life." *Nature* (September 13, 1974).

## CHAPTER NINE

Anderson, Alan. "The Mid-Atlantic Ridge: Window on the Earth's Core." *SR/World* (November 30, 1974).
Colbert, E. H. "Antarctic Fossils and the Reconstruction of Gondwanaland." *Natural History* (January 1972).
Coleman, P. J. "Seafloor Spreading Theory and the Odyssey of the Green Turtle." *Nature* (May 10, 1974).
"Continental Origins." *New Scientist* (June 12, 1975).
Decker, R. W. "Volcanism in Mexico and Central America." *Science* (March 16, 1973).
"Deep Sea Cores Indicate a Global Volcanic Spasm." *New Scientist* (February 1975).
Dietz, R. S., and J. C. Holden. "The Breakup of Pangaea." *Scientific American* (October 1970).
"Drilling Where the Continents Cracked." *New Scientist* (March 20, 1975).
Elders, W. A., and others. "Crustal Spreading in Southern California." *Science* (October 6, 1972).
"Evidence for Drifting Continents." *Chemistry* (April 1974).
Elliot, D. H., and others. "Triassic Tetrapods from Antarctica: Evidence for Continental Drift." *Science* (September 18, 1970).
Fooden, Jack. "Breakup of Pangaea and Isolation of Relict Mammals in Australia, South America and Madagascar." *Science* (February 25, 1972).
Ford, P. A. "Textbook Earthquake." *Environmental Journal* (November 1971).
Freed, W. K., and N. D. Watkins. "Volcanic Eruptions: Contribution to Magnetism in Deep-Sea Sediments Downwind from the Azores." *Science* (June 20, 1975).
Hallam, A. "Basin Tectonics." *Nature* (February 6, 1975).

———. "Alfred Wegener and the Hypothesis of Continental Drift." *Scientific American* (February 1975).
Kennett, J. P., and R. C. Thunell. "Global Increase in Quaternary Explosive Volcanism." *Science* (February 14, 1975).
Kurtén, Björn. "Continental Drift and Evolution." *Scientific American* (March 1969).
Orowan, Egon. "The Origin of the Oceanic Ridges." *Scientific American* (November 1969).
"The Peru Earthquake: A Special Study." *Bulletin of the Atomic Scientists* (October 1970).
Purrett, Louise. "Continental Drift and the Diversity of Species." *Science News* (December 11, 1971).
Smylie, D. E., and L. Mansinha. "The Rotation of the Earth." *Scientific American* (December 1971).
Sullivan, Walter. "Researchers See No Quake Limit." New York *Times* (December 25, 1975).
———. "Study of Pyramid Hints Shift on Earth." New York *Times* (February 28, 1973).
———. "Supercontinent Jigsaw: Last Piece Is Reported." New York *Times* (July 19, 1974).
"Volcanoes." *Chemistry* (October 1972).

## CHAPTER TEN

Alexander, Tom. "Ominous Changes in the World's Weather." *Fortune* (February 1974).
Broecker, W. S. "Floating Glacial Ice Caps in the Arctic Ocean." *Science* (June 13, 1975).
———. "Climatic Change: Are We on the Brink of a Pronounced Global Warning?" *Science* (August 8, 1975).
"Climatology According to the Greenland Ice Cap." *Science News* (May 10, 1975).
"The Cooling World." *Newsweek* (April 28, 1975).
Denton, G. H., and S. C. Porter. "Neoglaciation." *Scientific American* (June 1970).
Dresch, Jean. "Drought over Africa." *Development Forum* (May 1973).
Emiliani, Cesare. "Quaternary Paleotemperatures and the Duration of the High-Temperature Intervals." *Science* (October 27, 1972).

Francis, Peter. "Fire and Ice." *New Scientist* (July 3, 1975).
Harland, W. B., and M. J. S. Rudwick. "The Great Infra-Cambrian Ice Age." *Scientific American* (August 1964).
Hays, J. D., and others. "Variations in the Earth's Orbit: Pacemaker of the Ice Ages." *Science* (December 10, 1976).
"Intensive Climate Research Needed to Penetrate World's Cloudy Future." National Research Council press release (January 17, 1975).
Jaffe, Andrew. "Africa's Disastrous Drought." *Newsweek* (August 5, 1974).
Kukla, G. J., and R. K. Matthews. "When Will the Present Interglacial End?" *Science* (October 13, 1972).
Landsberg, H. E., and J. M. Albert. "The Summer of 1816 and Volcanism." *Weatherwise* (April 1974).
Likely, W. "2001: Hot or Cold?" *The Sciences* (January-February 1973).
"A 'Little Ice Age' in the North Atlantic?" *New Scientist* (March 13, 1975).
MacDonald, G. J. F. "The Modification of Planet Earth by Man." *Technology Review* (October-November 1969).
McCrea, W. H. "Ice Ages and the Galaxy." *Nature* (June 19, 1975).
Schmeck, H. M. "Climate Changes Called Ominous." New York *Times* (January 18, 1975).
———. "Ice Age Climate Studied with Help of Computers?" New York *Times* (May 12, 1975).
Sullivan, Walter. "Vast Antarctic Ice Sheet Studied for Clues to Periodic Ice Ages." New York *Times* (May 22, 1975).
"Sunsets and Stratospheric Dust Layers." American Institute of Physics press release (December 16, 1974).
Veeh, H. H., and John Chappell. "Astronomical Theory of Climatic Change: Support from New Guinea." *Science* (February 6, 1970).
"Volcanoes and Ice Ages: A Link." *Science News* (February 15, 1975).
Vuilleumier, B. S. "Pleistocene Changes in the Fauna and Flora of South America." *Science* (August 27, 1971).
Willis, Thayer. "Glaciers." *Science World* (November 29, 1971).

## CHAPTER ELEVEN

Commoner, Barry. "Nature Under Attack." *Columbia University Forum* (Spring 1968).
"Conserving Whales." *Science* (February 7, 1975).
"Death and Destruction Show the Path of Early Man." *New Scientist* (January 16, 1975).
Ehrlich, Paul. "Eco-Catastrophe." *Ramparts* (September 1969).
"Environmental Mutagenic Hazards." *Science* (February 14, 1975).
"Flopped Poles, Flipped Species." *Science World* (February 7, 1972).
"Following the Trail of a Magnetic Reversal." *Science News* (May 29, 1971).
"Genetic Vulnerability." *Agricultural Research* (July 1973).
"Genetic Vulnerability of Major Crops." National Research Council Study (August 1972).
Gould, Stephen. "The Great Dying." *Natural History* (October 1974).
Hallam, A. "Mass Extinctions in the Fossil Record." *Nature* (October 18, 1974).
Harlan, Jack R. "Our Vanishing Genetic Resources." *Science* (May 9, 1975).
Harrison, Gordon. "The Mess of Modern Man." *Natural History* (January 1970).
Hawley, Amos H. "Ecology and Population." *Science* (March 23, 1973).
Hays, James. "Reversals of the Earth's Magnetic Fields." *Bulletin of the Geological Society of America* (August 1972).
Henkin, Harmon, "Side Effects." *Environment* (January-February 1969).
Holden, David. "Egypt's Åswan Dam Isn't Doing the Job It Might Have." New York *Times* (May 11, 1975).
Libasi, Paul T. "Five Queasy Species." *The Sciences* (September 1974).
Long, A., and P. S. Martin. "Death of American Ground Sloths." *Science* (November 15, 1974).
"Magnetic Havoc." *Time* (November 30, 1970).
Martin, Paul S. "The Discovery of America." *Science* (March 9, 1973).

McVay, Scott. "The Last of the Great Whales." *Scientific American* (August 1966).
Mohr, Charles. "Tampering with Nature Perils Health." New York *Times* (April 20, 1975).
Newell, Norman D. "Crises in the History of Life." *Scientific American* (February 1963).
"Ocean Floor Record Links Dinosaurs, Plankton, Climate." *Science News* (October 23, 1971).
Paine, R. R., and T. M. Zaret. "Ecological Gambling." *Journal of the American Medical Association* (February 3, 1975).
Peakall, David G. "Pesticides and the Reproduction of Birds." *Scientific American* (April 1970).
Purrett, Louise. "When the North Pole Goes South." *Science News* (April 10, 1971).
———. "Magnetic Reversals and Biological Extinctions." *Science News* (October 30, 1971).
Rensberg, Boyce. "Elephants Are Declining Rapidly in Africa." New York *Times* (September 22, 1975).
"Strange Case of Dinosaur Eggs." *Chemistry* (October 1972).
Szent-Gyorgyi, Albert. "Snakes Do It. So Must Man." New York *Times* (March 29, 1975).
Uetz, George, and D. L. Johnson. "Breaking the Web." *Environment* (December 1974).
Webster, Bayard. "Butterflies to Be First Insects on U.S. Endangered List." New York *Times* (March 3, 1975).
———. "Plants Called Endangered, Along with Rare Animals." New York *Times* (November 7, 1975).

## CHAPTER TWELVE

"Analysis of Effects of Limited Nuclear Warfare." Committee on Foreign Relations, U. S. Senate (September 1975).
Griggs, D. T., and others. "Freon Consumption: Implications for Atmospheric Ozone." *Science* (February 14, 1975).
Haber, George. "The Crumbling Shield." *The Sciences* (December 1974).
Handler, Philip. "No Escape." New York *Times* (November 26, 1975).

Hester, N. E., and others. "Fluorocarbon Air Pollutants' Measurements in Lower Stratosphere." *Environmental Science and Technology* (September 9, 1975).

"Long-Term Worldwide Effects of Multiple Nuclear Weapons Detonations." National Research Council Study (October 1975).

McElheny, V. K. "The Future of Atom Power." New York *Times* (September 11, 1975).

"Mineral Resources and the Environment." National Research Council Study (February 1975).

Murdy, W. H. "Anthropocentrism: A Modern Version." *Science* (March 28, 1975).

"A Nuclear Horror." *Time* (February 9, 1976).

"Nuclear War: Federation Disputes Academy on How Bad Effects Would Be." *Science* (October 17, 1975).

"Nuclear War by 1999?" *Harvard Magazine* (November 1975).

"Rasmussen Issues Revised Odds on a Nuclear Catastrophe." *Science* (November 14, 1975).

# INDEX

"WARSHOFSKY SKEWERS THE SMUG NOTION THAT HISTORY IS ALL CONTINUITY."
— ALVIN TOFFLER

- The mathematics of catastrophe
- Black Holes— when Nature abandons the rules
- Ancient myths— true forecasts of coming cataclysm
- The chain of earthquakes that could crack open the Earth
- The astounding possibilities of biocatastrophe
- Velikovsky . . . Einstein . . . explanations established science prefers to dismiss

IS DESTRUCTION THE RULING FORCE OF THE UNIVERSE?

In DOOMSDAY: THE SCIENCE OF CATASTROPHE, two-time Emmy Award-winner Fred Warshofsky— whose credits include TV's "The 21st Century" and "In Search of Ancient Mysteries"— offers overwhelming evidence that it is!

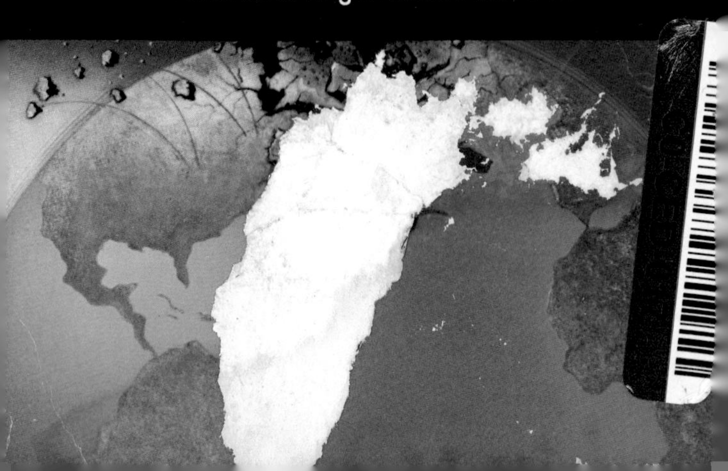